THE CAMBRIDGE
FIELD GUIDE TO
PREHISTORIC
LIFE

THE CAMBRIDGE FIELD GUIDE TO
PREHISTORIC LIFE

David Lambert
and the Diagram Group

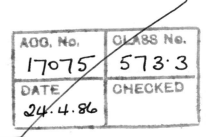

The right of the
University of Cambridge
to print and sell
all manner of books
was granted by
Henry VIII in 1534.
The University has printed

CAMBRIDGE UNIVERSITY PRESS

London   New York   New Rochelle   Melbourne   Sydney

Published by the Press Syndicate of the University of Cambridge
The Pitt Building, Trumpington Street, Cambridge, CB2 1RP
32 East 57th Street, New York, NY 10022, USA
10 Stamford Road, Oakleigh, Melbourne 3166, Australia

First published 1985

Printed in Great Britain at the University Press, Cambridge

British Library Cataloguing in Publication Data:

Lambert, David, *1932*–
   The Cambridge field guide to prehistoric life
   1. Palaeontology
   I. Title   II. Diagram Group
   560   QE711.2

ISBN 0 521 26685 8
ISBN 0 521 31299 X

The Diagram Group

Art director	Mark Evans
Artists	Graham Rosewarne, and
	Alastair Burnside, Brian Hewson, Richard Hummerstone,
	Karen Johnson, Alison Jones, Pavel Kostal,
	Jerry Watkiss
Art assistant	Robert Jones
Editor	Ruth Midgley
Indexer	Mary Ling

Consultants   Dr R. A. Fortey, British Museum (Natural History), London
Dr Angela Milner, British Museum (Natural History), London
Dr Ralph E. Molnar, Queensland Museum, Queensland, Australia
Professor R. J. G. Savage, University of Bristol, England
Mr C. A. Walker, British Museum (Natural History), London

Fossil hunters and fossils in
Wirksworth Cave, Derbyshire;
from an engraving of 1851.

FOREWORD

This book is a concise key to prehistoric animals, plants, and other organisms. It is the first to use field-guide techniques to picture and describe all the major forms of prehistoric life, giving precise details of what was what and what lived when. Large, labelled, life-like restorations, reconstructed skeletons, marginal "field-guide" illustrations, diagrams, and family trees – all integrated with the text – help readers to grasp important information at a glance. All this, and a use of both popular and scientific terms, will make the guide accessible to anyone from the inquiring eleven-year-old to the budding scientist.

There are ten chapters. Each has a brief explanatory introduction, followed by topics arranged under bold headings.

Chapter 1 (Fossil Clues to Prehistoric Life) briefly shows what fossils are and what they tell us of past life forms, their evolution, classification, and extinction.

Chapter 2 (Fossil Plants) describes briefly the major groups of plants known from the fossil record.

Chapter 3 (Fossil Invertebrates) ranges through all major groups of prehistoric animals that lacked a backbone.

Chapter 4 (Fossil Fishes) covers the world's earliest great group of backboned animals.

Chapter 5 (Fossil Amphibians) deals with the first group of backboned animals designed to walk on land.

Chapter 6 (Fossil Reptiles) describes vertebrates that dominated life on land for over two hundred million years.

Chapter 7 (Fossil Birds) gives a brief account of all known orders.

Chapter 8 (Fossil Mammals) gives examples from all major branches of the backboned group that replaced reptiles as masters of the land.

Chapter 9 (Records in the Rocks) shows which plants and creatures thrived through different periods and epochs on our ever-changing planet.

Chapter 10 (Fossil Hunting) tells how fossil hunters work and use what they discover. The chapter outlines achievements of famous palaeontologists and gives a worldwide list of fossil collections.

Lastly there is a list of books for further reading, and an index.

The producers of this guide consulted many works, but owe a special debt to these writers of books listed under Further Reading: U. Lehmann and G. Hillmer (invertebrates), and Edwin Colbert, Alan Feduccia, and Alfred Sherwood Romer (vertebrates). The author is answerable for all facts here presented, but thanks those named and unnamed experts whose advice has helped to make this book more accurate and up to date.

CONTENTS

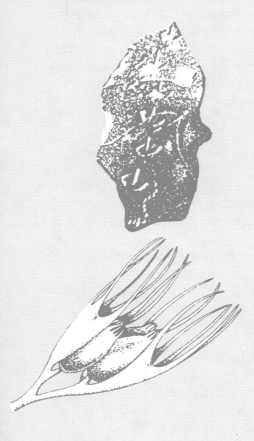

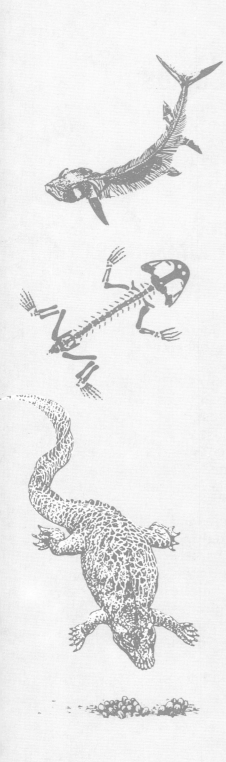

Chapter 4
FOSSIL FISHES

Chapter 5
FOSSIL AMPHIBIANS

Chapter 6
FOSSIL REPTILES

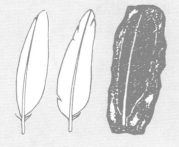

Chapter 7
FOSSIL BIRDS

Chapter 8
FOSSIL MAMMALS

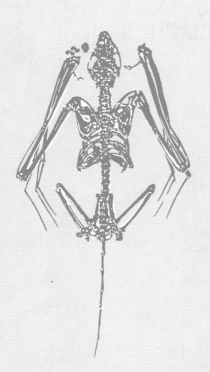

Chapter 9
RECORDS IN THE ROCKS

Chapter 10
FOSSIL HUNTING

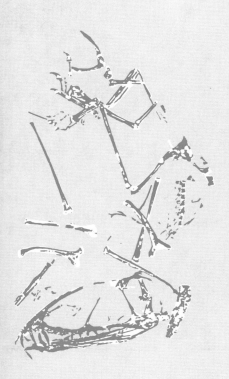

Chapter 1

This chapter explains fossils as keys that help us to unlock the puzzle of past life.

We show how parts of prehistoric organisms have survived as fossils in layered rocks, and what those fossils reveal – about the ages of the rocks themselves, the evolution (and extinction) of past life forms, and how and where prehistoric plants and creatures lived.

The chapter ends with a glimpse of Earth's oldest organisms – precursors of those plants and animals which, group by group, fill later pages of this book.

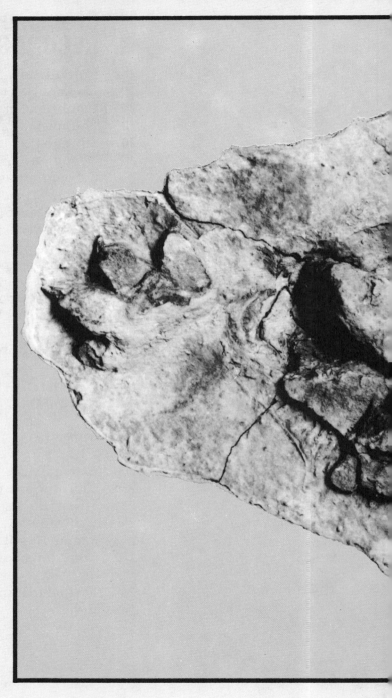

Fossil footprints tell us much about the shape, size, stance, and speed of otherwise unknown prehistoric animals. This print from Late Triassic rocks in Cheshire, England, came from *Cheirotherium* ("mammal hand"), a beast with a huge, thumb-like little finger. Narrow trackways show it walked on all fours, with limbs held well below the body. The feet would have resembled those of the thecodont reptile *Euparkeria*.

What fossils are

When plants or animals die, they usually decay. Sometimes, though, their hard parts get preserved in rock as fossils. Fossils are the clues that tell us what we know of long-dead living things.

Fossils form in several ways. The process usually happens under water. First, a newly dead plant or animal sinks to the bottom of a lake, sea, or river. Soft tissues soon rot. But before bone or wood decay, sand or mud may cover them, shutting out the oxygen needed by bacteria that cause decay. Later, water saturated with dissolved minerals seeps into tiny holes in the bone or wood. Inside these tiny pipes the water sheds some of its load of minerals. So layers of substances such as calcite, iron sulphide, opal, or quartz gradually fill the holes. This strengthens the bone or wood and helps it to survive the weight of sand or mud above. Sometimes bits of bone or wood dissolve, leaving hollows that preserve their shapes – fossils known as "moulds". If minerals fill a mould they form a "cast".

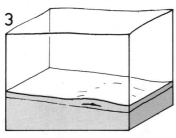

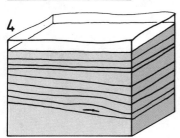

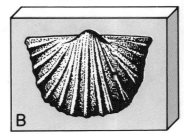

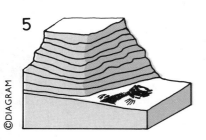

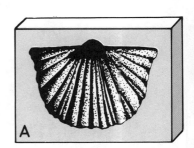

A fossil's story (left)
1 A fish that has just died lies on a sea bed.
2 Flesh rots, revealing bones.
3 Mud or sand covers the bones, preventing decay.
4 Layers of mud and sand bury the bones, now reinforced and fossilized by minerals.
5 Weather exposes the fossil bones by eroding the layered sediments above, long since hardened into stone and raised by uplift of the Earth's crust.

Mould and cast (right)
A A shell in rock dissolved to leave this shell-shaped hollow, called a mould.
B Minerals later filled the mould to form a cast.

Not only wood and bone become preserved. Skin, leaves, burrows, footprints, other tracks, and even droppings may form fossils. But soft-bodied creatures such as worms form fossils only in the finest fine-grained rock.

As a fossil hardens under water, layers of mud or sand grow above it. Their crushing weight and any natural cements that they contain may change thick layers of sand or soft mud into thin beds of hard rock. Millions of years later, great movements of the Earth's crust might heave up these beds to build mountains. Rain, frost, and running water slowly wear them down. In time, weather bares the mountains' inner layers and their fossils. Many fossils develop as we have described. But some form under sands piled up by desert winds, while amber, frozen mud, and tar preserve some ancient organisms whole.

Ant in amber
One hundred million years ago resin leaking from a tree trunk trapped this worker ant. The resin hardened into amber, preserving all but the ant's soft internal organs.

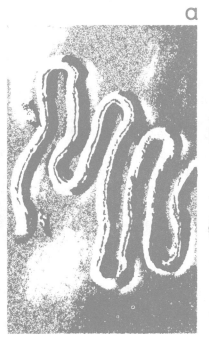

Fossil tracks
Two examples show different types of fossil tracks (not drawn to scale).
a Trail made probably by a snail-like creature, and preserved in Pennsylvanian (Late Carboniferous) rock.
b Footprints and beak marks left in mud by *Presbyornis*, an Eocene wading bird of North America. (Illustration after Erickson.)

15

Fossils in the rocks

Rocks of North America
a Precambrian rocks: older than 600 million years
b Palaeozoic rocks: 600–225 million years old
c Mesozoic rocks: 225–65 million years old
d Cenozoic rocks: less than 65 million years old

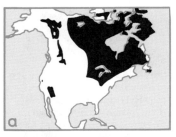

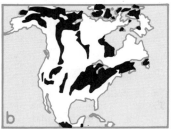

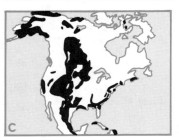

Fossils hold clues to the ages of the rocks containing them – sedimentary rocks such as limestones, mudstones, and sandstones. Each rock layer or "stratum" includes fossils special to that layer – remains of organisms living when the stratum formed. A cliff, cutting, or quarry may reveal many strata, like layers in a slice of cake. Usually the lowest layers are the oldest. Fossil organisms in these layers may have lived many million years before organisms in the top strata.

Geologists group rock strata in "systems" that give their names to geological "periods" in which the rocks formed. Groups of periods form larger time units called "eras". A period can be divided into Early, Middle, and Late, corresponding with the Lower, Middle, and Upper strata of the system that it represents. Some periods are divided into named "epochs", each corresponding with a "series" of rocks. And each series can be subdivided into "stages" or "formations".

Different kinds of prehistoric organisms lived at different times. Finding similar fossils in different continents helps geologists to correlate the ages of rocks around the world. This process is vital to the study called stratigraphy. Fossils show only the relative ages of the rocks. Radiometric dating gives rock ages more precisely. This method depends on radioactive elements inside some rocks. Half the substance in a radioactive element decays into another element in a known time called a half-life. In the next half-life, half of what remains decays, and so on. So certain key ingredients in ancient rocks reveal their ages to within a few million years. At first, radiometric dating worked well only with once-molten rocks in which there are no fossils. Now, radiometric techniques can help us date fossil-bearing sedimentary rocks as well.

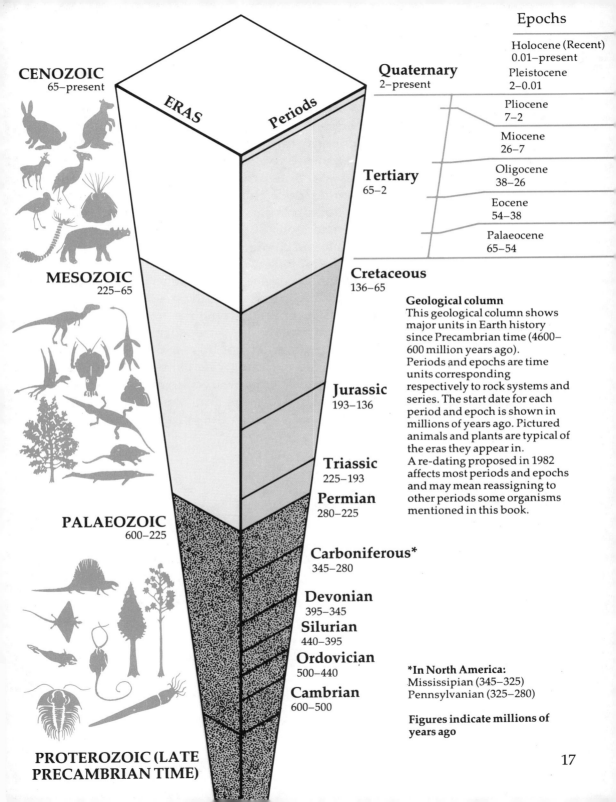

Epochs

Quaternary 2–present	Holocene (Recent) 0.01–present
	Pleistocene 2–0.01
Tertiary 65–2	Pliocene 7–2
	Miocene 26–7
	Oligocene 38–26
	Eocene 54–38
	Palaeocene 65–54

ERAS

Periods

CENOZOIC
65–present

MESOZOIC
225–65

PALAEOZOIC
600–225

PROTEROZOIC (LATE PRECAMBRIAN TIME)

Cretaceous
136–65

Jurassic
193–136

Triassic
225–193

Permian
280–225

Carboniferous*
345–280

Devonian
395–345

Silurian
440–395

Ordovician
500–440

Cambrian
600–500

Geological column
This geological column shows major units in Earth history since Precambrian time (4600–600 million years ago). Periods and epochs are time units corresponding respectively to rock systems and series. The start date for each period and epoch is shown in millions of years ago. Pictured animals and plants are typical of the eras they appear in. A re-dating proposed in 1982 affects most periods and epochs and may mean reassigning to other periods some organisms mentioned in this book.

*In North America:
Mississipian (345–325)
Pennsylvanian (325–280)

Figures indicate millions of years ago

17

Fossil clues to evolution

Fossils found in rocks of different ages show how living things evolved, or changed, through time. The first life forms were microscopically tiny organisms. Later came soft-bodied sea creatures. Some gave rise to animals with shells or inner skeletons. One group – the fishes – gave rise to amphibians. Amphibians led on to reptiles; reptiles separately to the birds and mammals. Body changes that produced each major group of organisms happened bit by bit. From one generation to another, slight shifts in inherited characters accumulated, in time producing brand new kinds of plant and animal.

Scientists can see such changes in the making by studying fossils in sequences of zones – biostratigraphic subdivisions of the rocks. "Key fossils" useful for this purpose include ammonites, brachiopods, and trilobites – fossil sea creatures widespread in rocks formed under ancient seas. (Most fossils were preserved in marine deposits.) Microfossils – fossil algae and other tiny fossils – are other valuable guides. So, too, are the minute fossil

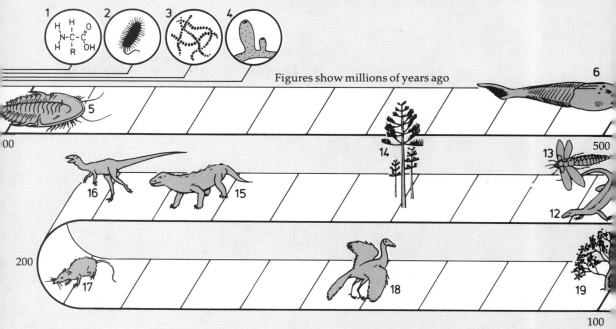

Figures show millions of years ago

18

spores and pollen grains produced by plants. Indeed, palynology – the study of fossil spores and pollen – is a special branch of fossil studies. Gaps blur the fossil record. Some organisms left no trace. Others have yet to be discovered. But enough remain for us to learn which organisms came from what – at least for many major groups. Wary palaeontologists watch out for homeomorphs: unrelated "lookalike" species similarly adapted for the same lifestyle.

Most major groups are many million years old. Within these, though, each kind of organism endured only as long as it could fend off enemies and rivals. New kinds of lethal enemy or harsh climatic changes wiped out unresistant species by the dozen. Fossils show that major evolutionary changes came in fits and starts. After mass extinctions (see pp. 28–29) new life forms sprang up and diversified explosively. New predators and herbivores soon populated habitats emptied of their old-established counterparts.

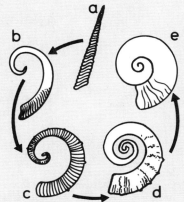

Evolving ammonites
These five fossils from successively younger rocks reveal one line of evolution among ammonites, molluscs with coiled shells.
a Shell straight
b Shell curved
c Shell strongly curved
d Shell loosely coiled
e Shell tightly coiled

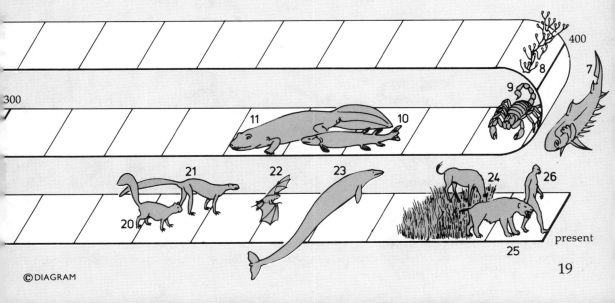

How living things are classified

Biologists classify or group all organisms, alive or extinct, according to how closely they resemble one another, or are related. Either way, those that can breed among themselves but not with others form one "species". Different species resembling one another more than they resemble other species form a "genus". Similar genera make up a "family".

Man's place in nature
Scientists classify man in these progressively higher categories and (not illustrated) subgroups.
1 Species: *Homo sapiens*
2 Genus: *Homo*
3 Family: Hominidae
Superfamily: Hominoidea
4 Order: Primates
5 Class: Mammalia
Subphylum: Vertebrata
6 Phylum: Chordata
7 Kingdom: Animalia
8 Superkingdom: Eukaryota

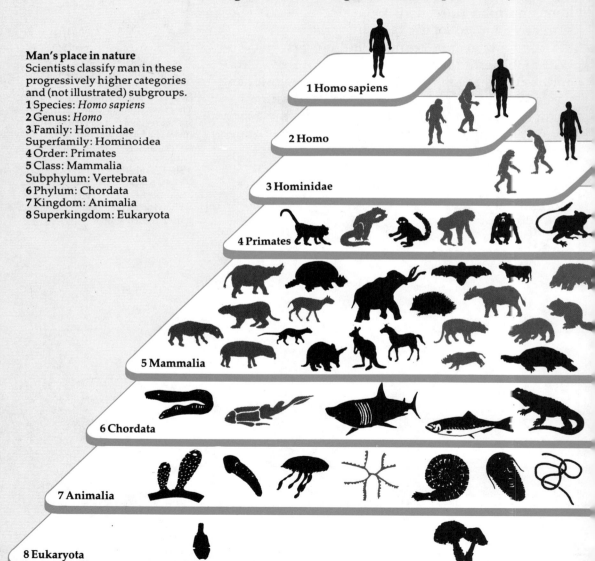

1 Homo sapiens
2 Homo
3 Hominidae
4 Primates
5 Mammalia
6 Chordata
7 Animalia
8 Eukaryota

Similar families form an "order"; similar orders make a "class"; similar classes make a "phylum" or, if plants, a "division". Similar phyla or divisions form a "kingdom".

Various experts disagree about how some organisms should be classified. "Lumpers" lump together in one group organisms that "splitters" would split up among several groups.

Scientific names of living things come at least partly from Greek or Latin. (Animal family names end in -idae, plant family names usually end in -ceae.) Scientists of all nations use the same scientific name for a given kind of animal or plant. This helps prevent confusion.

Man's fellow creatures
The diagram shows man's relationships to other living things. You could construct a similar diagram to represent the relationships of any other species.

Each level shows a grouping comprising lesser groups with the status of the group above. Individual creatures stand for groups. For instance, each creature shown for the phylum Chordata stands for a different class of animal. Each creature shown for the kingdom Animalia stands for a different phylum, and so on. Animals shown as pale shapes stand for groups that are extinct. Some groups are not depicted.
1 Species *Homo sapiens*
2 Genus *Homo*, containing three species
3 Family Hominidae, containing three genera
4 Order Primates, containing maybe 18 families
5 Class Mammalia, containing more than 30 orders
6 Phylum Chordata, containing more than a dozen classes, nine of which are vertebrates (backboned animals)
7 Kingdom Animalia, containing 19 phyla, all but one of them invertebrates (animals without a backbone)
8 Superkingdom Eukaryota, containing four kingdoms (animals, plants, fungi, and protists)

The tree of life

Scientists believe that all living things can be related to ancestors originating in minute one-celled organisms. First, lightning and ultraviolet radiation acting on a primeval atmosphere formed organic compounds from simple chemicals. Next, organic compounds organized in self-replicating committees formed simple, one-celled organisms – bacteria and blue-green algae – called collectively prokaryotes. Committees of prokaryotes arguably then produced more complex organisms – protists, fungi, plants, and animals – called collectively eukaryotes. If the superkingdom Prokaryota represents the main trunk of the evolutionary tree of life, Eukaryota – other living things – provides the branches.

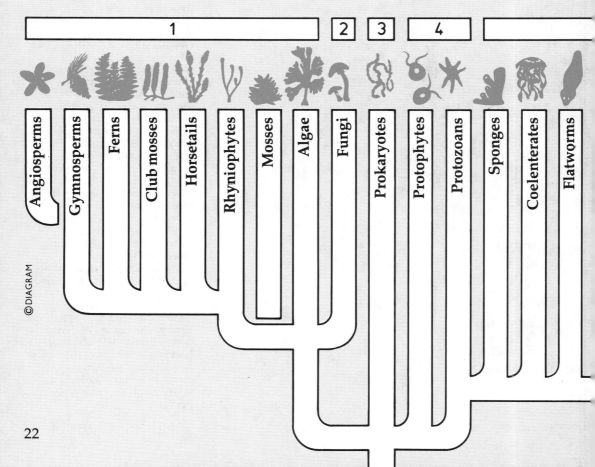

©DIAGRAM

22

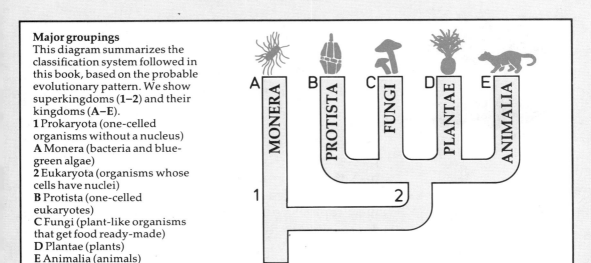

Major groupings

This diagram summarizes the classification system followed in this book, based on the probable evolutionary pattern. We show superkingdoms (**1–2**) and their kingdoms (**A–E**).

1 Prokaryota (one-celled organisms without a nucleus)

A Monera (bacteria and blue-green algae)

2 Eukaryota (organisms whose cells have nuclei)

B Protista (one-celled eukaryotes)

C Fungi (plant-like organisms that get food ready-made)

D Plantae (plants)

E Animalia (animals)

A MONERA
B PROTISTA
C FUNGI
D PLANTAE
E ANIMALIA

1

2

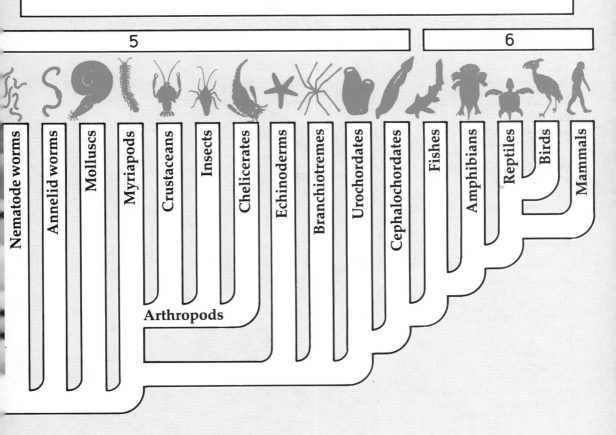

5

6

Nematode worms
Annelid worms
Molluscs
Myriapods
Crustaceans
Insects
Chelicerates
Echinoderms
Branchiotremes
Urochordates
Cephalochordates
Fishes
Amphibians
Reptiles
Birds
Mammals

Arthropods

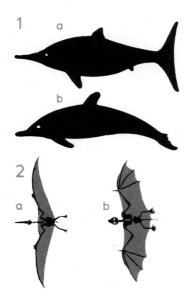

Convergent evolution (above)
Unrelated individuals evolved similar shapes suited for similar modes of life.
1 Fish-like swimmers:
a ichthyosaur (a reptile)
b dolphin (a mammal)
2 Skin-winged fliers:
a pterosaur (a reptile)
b bat (a mammal)

Death assemblage (right)
This diagram shows how the remains of living creatures can get mixed and broken after death to form a fossil death assemblage.
1A Prehistoric land and sea communities of organisms
1B Fossils from these sources and another source, mixed by erosion and undersea currents, and incorporated in rock
a Fossil derived from an older rock layer
b, c Remains of land plants and a land animal
d Soft-bodied creature that has left no fossil trace
e Shellfish shell halves separated and realigned
f Fragile crinoids broken up

How ancient organisms lived

Fossils reveal much about how individual organisms lived. Sometimes we can even learn how prehistoric plants and creatures interacted, as prey or predators.

Just by looking at a fossil creature, experts may be able to work out how it ate, moved, sensed, or even grew. Clues often lie in jaws or other features similar to those of certain living animals with habits that are known. For instance, prehistoric sea reptiles called placodonts had broad, flat crushing teeth like those of skates – fishes that crunch up sea-bed shellfish. It seems that placodonts ate shellfish too. Ichthyosaurs were reptiles with paddle-shaped flippers rather like a whale's. Plainly, ichthyosaurs, like whales, were accomplished swimmers, unable to walk on land. Trilobites were sea creatures resembling wood lice and with eyes rather like a fly's. Close study of one fossil trilobite proves that it saw in all directions, detecting other creatures' sizes and movements, though not their shapes. Fossils also show that growing trilobites shed their outer "skin" from time to time, to add an extra segment to the body. This

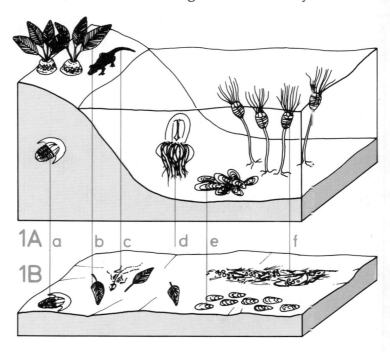

gives a rough idea of individuals' ages. (Other clues to fossil organisms' ages include tooth wear in mammals, and growth rings in tree trunks, fish scales, and mollusc shells.)

Where many different fossils crop up together in the same rock zone, palaeontologists may be able to identify plants, plant-eating animals, and the predators that preyed upon these herbivores. In this way an expert can work out a prehistoric food chain whose links consist of eaters and eaten – even perhaps a food web made up of interlinking chains.

Working out links is easy when fossils form a life assemblage – a group of organisms preserved as they once lived. Unfortunately many fossil groups are death assemblages. Such groups can include "outsiders" washed in by floods from other habitats. Then, too, predators, winds, currents, or chemicals destroy fragile bones and shells, or carry them away. So certain species from a given habitat survive only as broken scraps, or disappear completely.

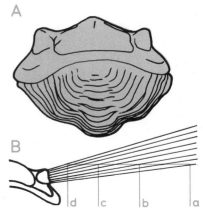

Trilobite abilities (above)
A This fossil trilobite was found curled up for protection.
B Study of the many lenses in its eyes shows that lenses at ever higher levels detected the advance (from **a** to **d**) of objects of a certain height.

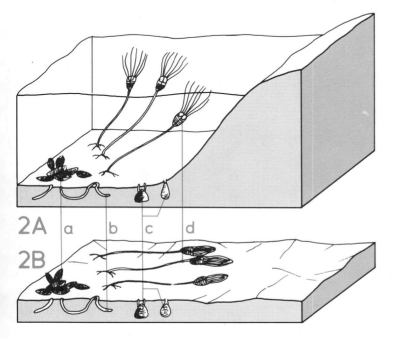

Life assemblage (left)
This shows how creatures' hard parts and tracks may be preserved intact in the positions where they lived.
2A Prehistoric sea-bed community of organisms
2B Fossils of the community undisturbed by erosion or undersea currents before being incorporated in rock
a Shellfish shells intact and aligned as during life
b Tracks of a burrowing worm preserved in rock
c Burrowing bivalve and sea urchin, with hollows their soft parts had made in sand
d Fragile crinoids unbroken

©DIAGRAM

25

What lived where?

Fossils tell us much about the surface of the world in ancient times. Then, as now, different species were designed for life in different habitats – for instance forest, grassland, desert, swamp, or river. Fossils therefore tell us indirectly about the kind of place they lived in – and about its climate. Widespread finds of fossil desert animals dating from a given time hint that desert climates were widespread too. Desert fossils crop up in Permian rocks as far apart as the United States, the USSR, and South Africa.

Fossils of lush tropical vegetation show that much of North America and Europe had a warm, wet climate in later Carboniferous times, when northern

Supercontinent (above)
This global view shows the supercontinent Pangaea in Permian times.

Worldwide life
Fossil finds of land organisms
(**1–2**) and the matching rims of continents (**A–F**) are clues to the extent and shape of Pangaea.
1 *Glossopteris*, a Permian tongue-shaped leaf, maybe from a tree-fern
2 *Lystrosaurus*, an Early Triassic mammal-like reptile
A North America
B South America
C Africa
D India
E Antarctica
F Australia

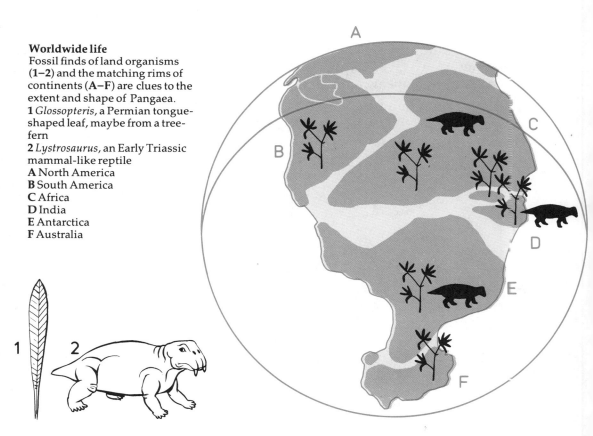

lands lay close to the Equator. Fossils have done much to prove that continents have drifted from their old positions. For instance, the Permian fossil plant *Glossopteris* occurs in all southern continents, now widely separated by oceans. Plainly, when *Glossopteris* grew, all southern continents lay locked together. Geologists believe they have been tugged apart by currents in the molten rock beneath the oceans, where Earth's crust of solid rock is thin and weak. But finds of certain fossil species only in specific regions suggest that natural barriers such as mountains, seas, or temperature boundaries stopped those organisms spreading.

Isolated life (below)
Shown on this map (**1–8**) are mammals that evolved in isolation after Pangaea broke up and continents (**A–F**) drifted apart, some split by seas.
1 Uintatheres
2 Opossum rats
3 Pyrotheres
4 Aardvarks
5 Embrithopods
6 Lemurs
7 Insectivores
8 Spiny anteaters
A North America
B South America
C Africa
D Europe
E Asia
F Australia

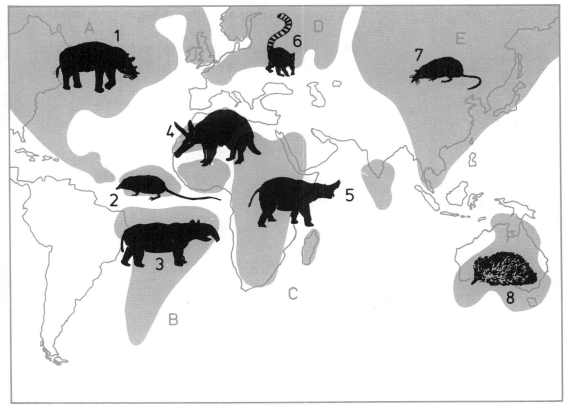

© DIAGRAM

Mass extinctions

Plants and animals alive today account for most known species. Yet species that became extinct must have outnumbered these by far. One calculation suggests there are 4.5 million living species, but that 980 million species evolved in the last 600 million years. This estimate supposes each species lasted on average for only 2.7 million years.

In fact some forms were far more durable. The Australian lungfish may have survived for about 100 million years. Sea creatures like the king "crab", coelacanth (a fish), and *Neopilina* (a mollusc) are close kin of beasts with even longer histories.

Somehow, such living fossils resisted changes that wiped out many other organisms. Sometimes, disasters struck down many groups together. The greatest mass deaths marked the ends of eras. Thus many sea creatures and the great group of reptiles called pelycosaurs died out as the Palaeozoic Era ended. Major reptile groups including dinosaurs, and those once-abundant molluscs ammonites, all vanished as the Mesozoic Era closed.

Lifelines
Thick and thin horizontal bands show the expansion and collapse of certain groups of organisms through prehistoric time. (Band thicknesses are not to scale.) Thick vertical lines show mass collapses changing the make-up of marine communities. Widespread land and sea extinctions marked the end of the Cretaceous period.
1 Archaeocyathids
2 Reef-forming stromatolites
3 Tabulate corals
4 Trilobites
5 Bivalves
6 Ammonites
7 Archosaurian reptiles (including dinosaurs)

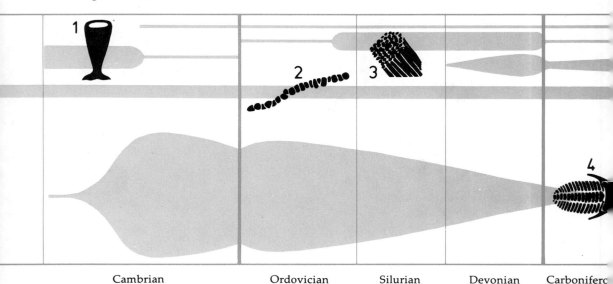

Cambrian   Ordovician   Silurian   Devonian   Carbonifero

Experts disagree about what caused these mass extinctions. Perhaps cosmic rays from an exploding star deformed unborn young. Maybe deadly rays from space poured down when the Earth's magnetic field reversed. Or perhaps changes in the positions and levels of the continents caused killing periods of cold.

Some scientists blame huge lumps of rock that could have crashed on Earth from space like massive bombs. The impact of a rock 10 kilometres across would hurl enough dust and moisture into the air to darken skies around the world for months. Plants and plant-eating creatures would die in droves. The world could briefly freeze. But once dust settled, moisture still up in the sky would trap the Sun's incoming heat. Then overheating could destroy creatures unable to control their body temperature.

Yet no mass extinction theory convincingly explains why many groups of living things survived.

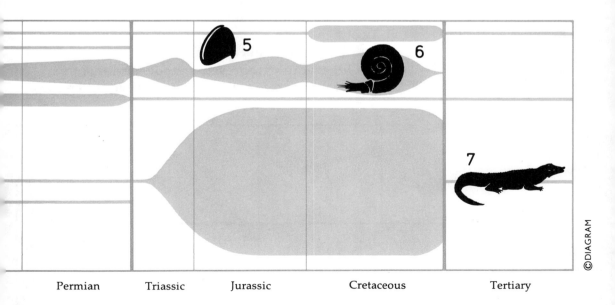

Permian   Triassic   Jurassic   Cretaceous   Tertiary

©DIAGRAM

The oldest fossils

The earliest identifiable living things were bacteria and blue-green algae – members of the superkingdom Prokaryota, tiny one-celled organisms whose genetic blueprints did not lie in a single nucleus. Prokaryotes probably evolved more than 3600 million years ago from combined amino acids created in the early ocean – a chemical-rich "test tube" shrouded by a methane-rich atmosphere intensively bombarded by solar radiation and electric storms.

Pioneer prokaryotes consumed ready-made amino acids. Later came blue-green algae containing the green pigment chlorophyll, which enabled them to use the energy in sunshine to build their own food compounds from carbon dioxide and water. This process, photosynthesis, yielded free oxygen as waste.

In time, free oxygen screened the Earth's surface from the Sun's harmful ultraviolet rays and formed a rich new source of energy. This was tapped by new, more complex life forms, members of the only other superkingdom: Eukaryota. Eukaryotes, with a central nucleus to each cell, comprise the fungi, protists, plants, and animals.

Our examples represent the sole prokaryote kingdom and one of the most primitive eukaryote kingdoms.

Accelerating evolution
Prokaryotes played a crucial role in evolution, as we show here. The thicker the line, the more diversified the life forms that it represents.
A Chemical evolution: amino acids, proteins, sugars, etc formed in water.
B Slow organic evolution: photosynthesizing prokaryotes enriched the atmosphere with oxygen; some formed ozone.
C Rapid evolution, mainly of eukaryotes deriving energy from oxygen and screened from harmful solar rays by ozone in the atmosphere. Evolution accelerated 1000–700 million years ago as atmospheric oxygen reached more than 1% of its present level.

©DIAGRAM

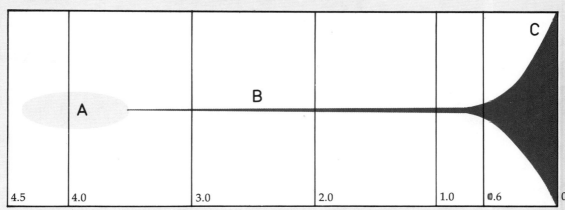

| 4.5 | 4.0 | 3.0 | 2.0 | 1.0 | 0.6 | 0 |

billion years ago

1 **Kakabekia** resembled a microscopic umbrella with a clubbed stalk. It might have been a bacterium dividing into two. Size: most bacteria are under 1 micron (0.001mm) across. Time: Precambrian (2000 million years ago). Place: Canada. Kingdom: Monera (bacteria and blue-green algae). Subkingdom: Bacteria (probably the earliest, most primitive living organisms).

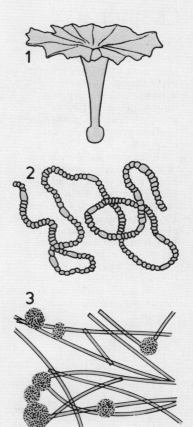

2 **Nostoc,** a blue-green alga, forms a necklace-like mass. It probably evolved from one-celled Precambrian ancestors. Kingdom: Monera. Subkingdom: Cyanophyta (blue-green algae, mostly tiny plants under 25 microns across).

3 **Aspergillus** is a mould, producing mildew on fruit, leather, or walls. Its growing, thread-like hyphae produce "side shoots" bearing spores from which new moulds grow. Size: minute. Kingdom: Fungi (plant-like organisms that lack chlorophyll and feed on dead or living organisms). Fossil hyphae have been found in Precambrian rocks, but claims for a 3800-million-year-old fossil fungus found in Greenland proved mistaken.

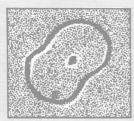
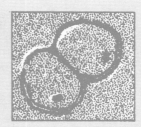

Early complex cells

Fossils of cells multiplying by division hint at early complex life 1000 million years ago. Shown here much magnified are one-celled eukaryotic algae fossilized in chert at various stages of division. Genes within cell nuclei determine how each daughter cell develops. Most multicellular organisms undergo sexual reproduction, in which a fertile egg cell receives genes from both parents. This innovation increased the chance of genetic variation and so of evolution.

31

Chapter 2

FOSSIL PLANTS

These pages give a brief evolutionary overview of the plant kingdom (Plantae) – that vast group of organisms on which all animals depend directly or indirectly for their food. We describe key features of major groups, and give examples.

The chapter starts with divisions of those primitive plants collectively called algae. For convenience we include protophytes: tiny, one-celled algae now usually ranked (with protozoans) in a separate kingdom from true plants (see pp. 20–23 for classification).

Most of the chapter deals in turn with increasingly advanced divisions, of which some of the names and groupings are disputed.

This artist's reconstruction shows a coal forest of Late Carboniferous (Pennsylvanian) Europe or North America. Huge horsetails, scale trees, ferns and tree ferns sprout from rich swamp mud and jostle one another for a share of light.

Simple plants

True plants (members of the kingdom Plantae) consist of cells surrounded by a cellulose cell wall, not just a membrane like that around animal cells. Plants contain the green pigment chlorophyll, enabling them to manufacture food from carbon dioxide and water.

The simplest plants are green, red and brown algae. (Other algae are classed as prokaryotes and protists.) Most green algae are freshwater plants. Green, red and brown algae comprise the seaweeds. The simplest kinds of land plants are bryophytes, including liverworts and mosses. Some have leaves but none has roots. All reproduce by spores. "True" algae evolved (perhaps from blue-green algae) at least 1000 million years ago. Green algae were probably ancestral to all higher plants. Bryophytes appeared at least 350 million years ago.

A brown alga
Here we show features of the inter-tidal seaweed *Fucus vesiculosus* (bladder wrack).
a Receptacles (swollen branch tips containing mucilage-filled conceptacles where sex cells develop)
b Air bladder, for buoyancy
c Blade
d Stipe
e Holdfast

Sexual reproduction
Fucus vesiculosus has the life cycle shown in this diagram.
A Male conceptacle
B Female conceptacle
1 Antheridium
2 Male sex cells being released
3 Oogonium
4 Female sex cells being released
5 Male sex cells surrounding a female sex cell – one male cell fertilizes the female

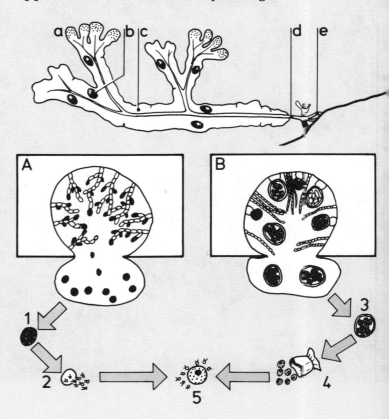

1 **Stephanochara,** an Oligocene stonewort, produced this female sex organ with a wall composed of helically arranged cells, shown magnified about 20 times. Division: Charophyta, an algal group that dates back to Devonian times.

2 **Ulva,** "sea lettuce", is a green, shallow-water seaweed. Size: 10–46cm (4–18in). Division: Chlorophyta (green algae), known since at least as early as Ordovician times.

3 **Fucus** is a brown shore seaweed. Height: 13–91cm (5in–3ft). Division: Phaeophyta (brown algae), perhaps dating from Devonian times.

4 **Ceramium** is a red seaweed, banded and with forked, inturned tips. Height: 2.5–30.5cm (1in–1ft). Division: Rhodophyta (red algae), dating from Ordovician times.

5 **Marchantia** is a liverwort. Size: 5cm (2in). Division: Bryophyta (liverworts and mosses), dating from Devonian times.

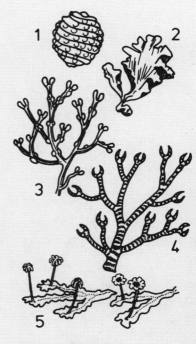

Colour and depth (left)
a–f Green seaweeds need red light which only penetrates shallow water.
g–l Brown seaweeds, whose brown pigments mask their green chlorophyll, live in inter-tidal zones.
m–o Red seaweeds, with a masking red pigment, can use blue light, which penetrates 100m (330ft) deep.

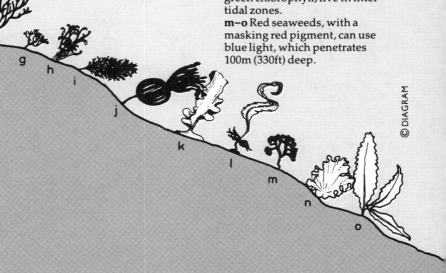

Protophytes

Described here are certain one-celled algae that possess a nucleus and can move about like animals. Scientists now usually rank such algae in the subkingdom Protophyta ("first plants") of the kingdom Protista. Protists evolved by 1200 million years ago, and probably came from simple cells that swallowed bacteria and even blue-green algae which then lived on inside them. The new "committee" micro-organisms reproduced by dividing, like bacteria, but most of their ingredients split too, in complex ways. Of our protophyte examples, number 1 is in the phylum Pyrrophyta (algae with yellowish or brownish food stores). Numbers 2–4 are in the Chrysophyta (algae with golden-brown food stores). Prehistoric protophytes mostly drifted in the sea. Their fossils help scientists date rocks formed in the last 200 million years.

1 **Gonyaulacysta** is known only from fossils of its resting form: a hard-walled cyst. Class: Dinophyceae (dinoflagellates), sea-surface micro-organisms

Food factory
This diagram shows (much magnified) structures in a tiny, living, one-celled protophyte, *Prymnesium*. Even such microscopic scraps of life are highly organized.
a Haptonema
b Flagellum (a whip-like structure used in locomotion)
c Golgi body (rich in fat)
d Pyrenoid (a protein body)
e Chloroplast (food-producing unit containing chlorophyll)
f Fat (stored energy supply)
g Nucleus (control centre essential for the cell's life and reproduction)
h Leucosin vesicle, containing food reserves

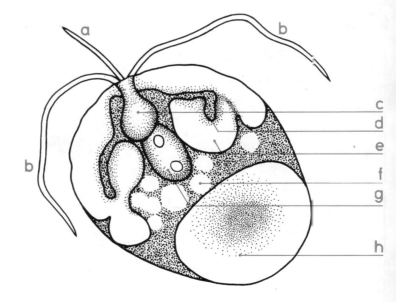

0.005–2mm across. Each has a groove around the
body and lashes itself along with two cilia ("whips").
Time: Silurian onward (guide fossils since Triassic
times).

2 A silicoflagellate These have one or two "whips"
and a tube- or rod-like opal skeleton, often fossilized.
Silicoflagellates are organisms 0.02–0.1mm across,
dating from Cretaceous times.

3 Cymbella has a two-valved skeleton like a glass
box with an overlapping lid. Length: 0.03mm. Time:
Pliocene onward. Class: Bacillariophyceae (diatoms)
– oblong, round, and other micro-organisms dating
from Cretaceous times. Accumulations of their shells
form diatomaceous earth.

4 Coccolithus is a ball-shaped sea micro-organism
covered in limy plates. Time: Pliocene onward.
Class: Coccolithophyceae – important rock formers
since Jurassic times. Their plates 0.002–0.01mm
across sink on death, building layers of chalk or
chalky mud.

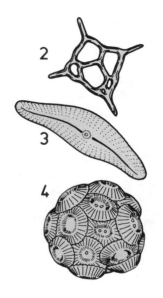

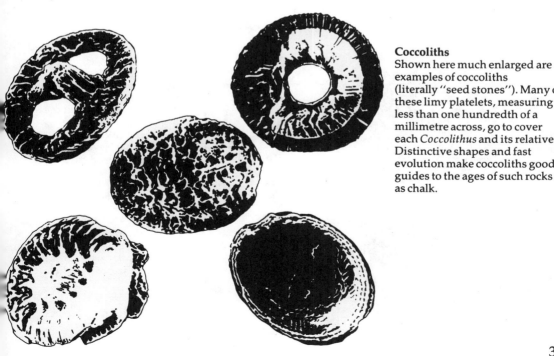

Coccoliths
Shown here much enlarged are
examples of coccoliths
(literally "seed stones"). Many of
these limy platelets, measuring
less than one hundredth of a
millimetre across, go to cover
each *Coccolithus* and its relatives.
Distinctive shapes and fast
evolution make coccoliths good
guides to the ages of such rocks
as chalk.

©DIAGRAM

Vascular plants

Vascular plants – plants with internal channels for transporting liquids – include ferns, conifers, and flowering plants. Algal ancestors gained support and nourishment in water but vascular plants evolved for life on land. Roots give anchorage and obtain water from soil. Stems raise the food-producing leaves up to the light. A waterproof cuticle prevents desiccation. Stomata – holes that can be closed – let gases in and out for food production and respiration. Above all, a "plumbing" system of tiny internal tubes transports water and salts up from the soil and dissolved food down from the leaves.

Vascular plants appeared about 400 million years ago. Pioneers had short, bare stems and neither roots nor leaves. Tall plants with woody stems became the first true trees. Among these were the giant club mosses and horsetails of swampy Carboniferous forests. Their compressed carbonaceous remains formed coal.

Here we describe prehistoric examples from three spore-producing divisions: Rhyniophyta, Lycophyta, and Sphenophyta.

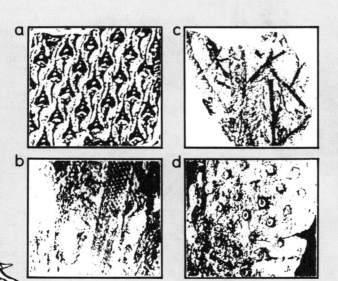

Lepidodendron restored
Fossil finds (left) enabled scientists to reconstruct the tree they came from (far left).
a Piece of trunk scarred by shedding old leaves
b Piece of grooved, mid-level trunk
c Branches with leaves
d Surface of root-like anchoring structure
e Whole tree, with a man to show its size. (Many other plant fossils left such fragmentary remains that we cannot ever guess the shape and size of the whole plant.)

1 **Cooksonia,** the first known vascular plant, had simply forked, leafless stems, each ending in a spore-filled cap. Height: about 5cm (2in). Time: Late Silurian. Place: Wales. Division: Rhyniophyta.

2 **Asteroxylon** had a main stem with forked branches bearing tightly packed, tiny leaflets. Height: up to 1m (3ft 3in). Time: Early Devonian. Place: Scotland. Division: Lycophyta (club mosses).

3 **Lepidodendron** was a giant club moss with a root-like anchoring organ divided into four main branches; a broad, tall, bare trunk crowned by two sets of repeatedly forked branches bearing narrow leaves; and small and large spores – the latter borne in cone-like structures. Height: 30m (100ft). Time: Carboniferous. Place: Europe and North America. Division: Lycophyta.

4 **Calamites** had a tall, jointed stem and upswept branches bearing rings of narrow leaves. Height: up to 30m (100ft). Time: Carboniferous. Place: Europe and North America. Division: Sphenophyta (horsetails).

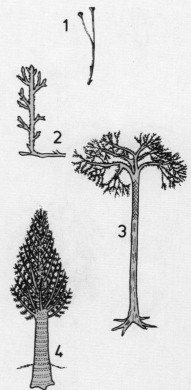

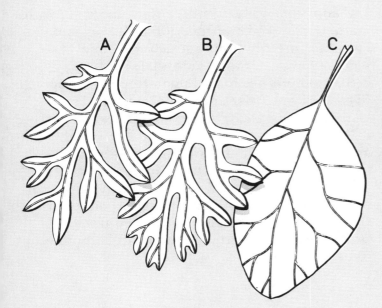

Leaves evolving *(left)*
These three illustrations retrace the likely evolution of compound leaves from stems.
A Branching stems of certain early plants probably had broad flat rims. These increased the surface area exposed to sunlight. Such stems trapped more light than others and so produced more food.
B Later plants produced more branches than their ancestors, so increasing the surface area still further.
C Later still came plants whose many, wide-rimmed branches fused to form broad plates that faced the Sun. These plates were leaves.

©DIAGRAM

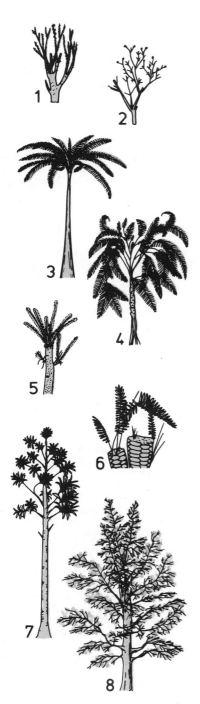

Ferns and gymnosperms

Ferns and gymnosperms probably arose from plants like *Cooksonia* more than 350 million years ago. Many ferns have feathery or strap-like leaves called fronds. They lack flowers or seeds and disperse by spores shed from the fronds' undersides. A spore falling on damp soil grows into a minute "breeding" plant producing male and female cells. Male cells fertilize female cells which then develop into normal fern plants.

Gymnosperms ("naked-seed" plants) include seed-ferns and conifers – flowerless plants that reproduce by seeds not spores. Seeds develop when wind-blown pollen fertilizes egg cells still on the parent plant. The ripe seeds have food supplies and waterproof coats. So fallen seeds survive drought, sprouting when rain soaks the soil. Ferns and seed-ferns flourished in warm, wet Carboniferous forests. Gymnosperms became increasingly varied during Mesozoic time. These pages show examples of the division Polypodiophyta or ferns (items 1–3), and gymnosperms, an artificial grouping (items 4–9).

1 **Cladoxylon** was a primitive "fern ancestor" with a main stem, forked branches, forked leaves, and fan-like spore-containing structures. Time Mid-Late Devonian. Order: Cladoxylales (extinct).
2 **Stauropteris** had forked stems tipped with spore-filled caps, much like ancestral *Cooksonia*. Time: Carboniferous. Order: Coenopteridales (extinct).
3 **Psaronius** was a prehistoric tree fern. Height: about 7m (23ft). Time: Carboniferous-Permian. Order: Marattiales (still surviving in the Tropics).
4 **Medullosa** was a seed-fern: producing seeds not spores. Height: 4.6m (15ft). Time: Carboniferous. Order: "Pteridospermales" – the seed-ferns, primitive gymnosperms abundant in Carboniferous forests.

5 **Williamsonia** sprouted stiff, palm-like fronds from a trunk with scars where stems of old, dead leaves had fallen off. Height: 3m (10ft). Time: Jurassic. Order: Bennettitales (once widespread but long extinct).

6 **Zamia** has leathery fronds rising from a short, fat, pithy stem. It is a survivor of the abundant Mesozoic order Cycadales (cycads).

7 **Cordaites** had a tall, slim, straight main trunk with a crown of branches bearing long, strap-shaped leaves. Seeds grew on stalks from cone-like buds. Height: up to 30m (100ft). Time: Carboniferous. Order: Cordaitales (long extinct).

8 **Ginkgo,** a living fossil, has fan-shaped leaves. Height: up to 30m (100ft). Time: Permian onwards, widespread in the Jurassic. Order: Ginkgoales ("maidenhair trees").

9 **Araucaria** (monkey puzzle) is an evergreen, cone-bearing tree with stiff, flat, pointed leaves. Height: up to 45m (150ft). Time: Jurassic onwards. Order: Coniferales (conifers – pines, firs, etc).

9

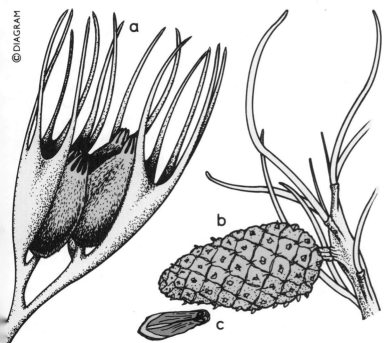

a

b

c

Seeds and cones
These two examples contrast the seeds and seed-containing structures of an early and a modern gymnosperm.
a Forked branches form the "bracts" which only loosely surround the seeds in a Late Devonian plant called *Archaeosperma*.
b A woody cone surrounds many individual seeds in a modern conifer. The cone opens when the seeds are ripe, to let them fall.
c A conifer seed has a built-in, wing-like blade. Ripe seeds flutter lightly down, dispersing on the wind.

© DIAGRAM

Flowering plants

Flowering plants are among the most successful land plants ever. They thrive from high latitudes to the Equator, from seashore to mountaintop. They include most garden plants, farm crops, and broadleaved trees, and range in size from 100-metre (330ft) *Eucalyptus* trees to tiny duckweeds.

Reasons for success included a carpel, a protective covering that earns this major group of vascular plants its scientific name: Angiospermae ("enclosed seed plants"). Angiosperm seeds develop in greater safety than the "naked" seeds of gymnosperms, their likely ancestors. Finds of fossil leaves and pollen hint that flowering plants evolved about 120 million years ago. By 65 million years ago over 90 per cent of known fossil plants were angiosperms. There were beeches, birches, maples, poplars, walnuts and many other familiar kinds. Since Tertiary times the main change has been in distribution, as warmth-loving species tended to retreat from cooling polar regions.

Our examples represent both subclasses: the Monocotyledonae (monocots, with one seed-leaf) and Dicotyledonae (dicots, with two seed-leaves). Dicots form the vast majority of angiosperms.

1 **Buchloe,** the living buffalo grass, represents the grasses – monocots that first became plentiful in the Miocene. Order: Graminales (grasses and sedges).

2 **Magnolia** is a genus of dicot trees and shrubs with large leaves and showy flowers regarded as primitive in structure. Time: Early Cretaceous onwards. Place: Asia and the Americas. Order: Ranales, among the earliest of all known flowering plants.

1

2

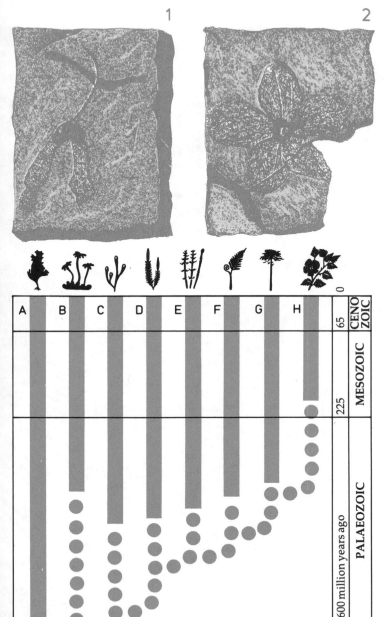

Fossil angiosperms
These Miocene plant fossil remains come from Switzerland. Fossil seeds and flowers are far rarer than fossil leaves but most fossil angiosperms closely resemble living kinds.
1 Winged seeds of a Miocene maple, *Acer trilobatum*
2 The blossom of *Porana oeningensis*

Plants family tree
This shows important groups of true plants. Groups B, C, D, E, F and H represent divisions. A and G are artificial groupings.
A Algae (four divisions, some older than others)
B Bryophyta (liverworts and mosses)
C Rhyniophyta (*Cooksonia*)
D Lycophyta (club mosses)
E Sphenophyta (horsetails)
F Polypodiophyta (ferns)
G Gymnosperms
H Angiospermae (flowering plants)

43

Chapter 3

FOSSIL INVERTEBRATES

Most phyla (the major groups of animals) are invertebrates – creatures with no backbone. According to some experts, invertebrates account for 18 of 19 phyla comprising the kingdom Animalia. (Even the nineteenth phylum includes some groups without a backbone.) This chapter surveys the chief invertebrate groups significant as fossils, from simple sponges to complex echinoderms. For convenience, we start with the tiny protozoans, now usually ranked (with protophytes) in a separate kingdom from true animals (see pp. 20–23 for classification).

Early, soft-bodied invertebrates left few fossil traces. But most phyla may have appeared in the sea between 800 and 600 million years ago. This upsurge in evolution became possible as Earth's atmosphere became breathable, and ozone shielded the surface of the sea from lethal ultraviolet radiation.

All invertebrate phyla that evolved survive today, but through the ages many subgroups disappeared.

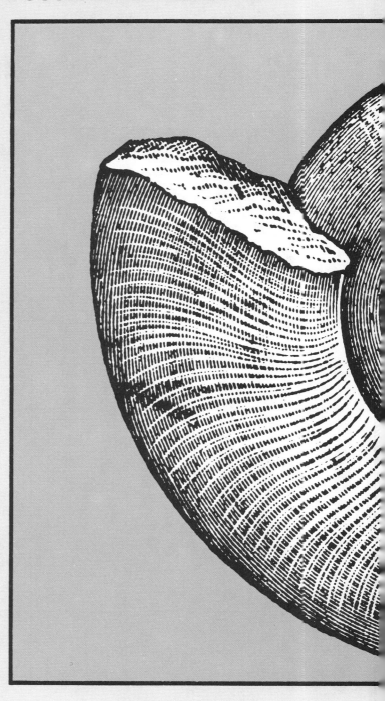

Clymenia sedgwickii here represents that huge, long-lived group of invertebrates, the ammonoid cephalopods. Part of its outer shell has been removed to show the distinctive suture lines of its internal chambers. Sutures help to identify this as a goniatite ammonoid. Clymeniids thrived in Late Devonian seas. (The Mansell Collection.)

About invertebrates

Invertebrates is a name popularly used for the larger, lowlier section of the animal kingdom. This huge group – including insects, worms, snails, etc – comprises living things with a distinct body shape and size that develop from an embryo or larva formed when male and female sex cells meet. Most animals move about and, unlike plants, all feed upon organic matter, for they cannot manufacture food. Zoologists put animals in two subkingdoms: the primitive Parazoa (sponges and close kin), and the more advanced Metazoa (animals composed of many cells of different types, specialized for performing different tasks). All parazoans and most metazoans are invertebrate: they lack a notochord ("backstring") or vertebrae (bones making up a backbone). For convenience, this chapter includes those tiny one-celled organisms protozoans, although these rank as a subkingdom in the Protista, a kingdom that also includes microscopic plant-like organisms (see pages 36–37).

Borehole tracks from rocks in Zambia suggest that protozoans had given rise to parazoans and metazoans over 1000 million years ago. By 700 million years ago, jellyfishes, worms, and other lowly metazoans lived in the sea. This chapter traces different groups of fossil invertebrates, from minute protozoans through the progressively more complex sponges, corals, molluscs, worms, and arthropods, on to echinoderms – indirect ancestors of backboned animals. Most major invertebrate groups evolved in Palaeozoic times, yet still endure. So living species often help us to deduce what their fossil forebears looked like, and how these lived and moved about.

Invertebrate groups
This shows most major groups significant as fossils and featured in this chapter. Numbers indicate phyla, letters, subphyla.

1 Porifera (sponges)
2 Archaeocyatha (archaeocyathines)
3 Cnidaria (jellyfishes and corals)
4 Rhizopoda (rhizopods)
5 Ciliata (ciliates)
6 Aschelminthes (roundworms)
7 Annelida (segmented worms)

8 Arthropoda (arthropods)
a Onychophora (velvet worms)
b Trilobitomorpha (trilobites)
c Chelicerata (chelicerates – scorpions, spiders etc)
d Myriapoda (myriapods – millipedes etc)
e Crustacea (crustaceans)
f Insecta (insects)
9 Mollusca (molluscs)
g Amphineura (chitons)
h Conchifera (snails etc)
10 Phoronida (phoronids)
11 Bryozoa (bryozoans)

12 Brachiopoda (brachiopods)
13 Conodontophorida (conodonts)
14 Echinodermata
i Homalozoa (carpoids)
j Blastozoa (blastoids etc)
k Crinoidea (sea lilies)
l Asterozoa (sea anemones)
m Echinozoa (sea urchins etc)
15 Branchiotremata (branchiotremes)

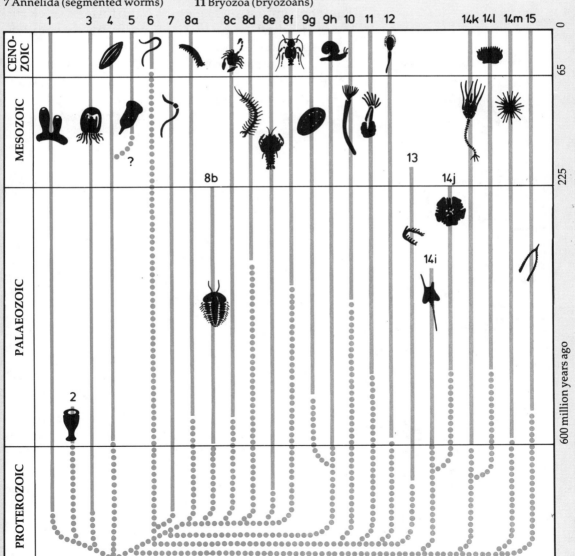

Protozoans

Protozoans ("first animals") form a subkingdom in the kingdom of microscopic one-celled organisms, Protista. Unlike their plant-like counterparts (see pages 36–37), protozoans tend to feed on other organisms or organic substances.

Scientists group protozoans in different phyla according to how they move about. Only prehistoric kinds with hard parts have survived as fossils. The earliest date back over 700 million years. Billions lived, and live, on sea floors or in surface waters. Their accumulating skeletons have formed thick rock layers in which some fossil types indirectly show the saltiness and warmth of ancient seas.

Of our examples, numbers 1–3 belong to the phylum Rhizopoda; number 4 to the phylum Ciliata.
1 **Fusulina** had a large, chalky, spindle-shaped skeleton tapered at both ends. Size: 6cm (2.4in). Time: Late Carboniferous. Class: Foraminifera

1

Fusulinid features
Here the big foraminiferan
Fusulina is shown enlarged.
a Large test (skeleton)
b Spindle (sometimes spherical) shape
c Composition: calcium carbonate
d Three- or four-layered spirotheca (wall)
e Concentric chambers
f Fluted septa (divisions)
g Septal pores
h Foramen (opening)

(foraminiferans – micro-organisms usually known as fossils from their calcareous shells pierced by tiny holes. From these, in life, thread-like "false feet" projected for locomotion and seizing particles of food). Order: Fusulinida (fusulines).

2 **Nummulites** belonged to a group of giant coin-shaped foraminiferans with limy, perforated shells comprising many chambers. Size: 1–6cm (0.4–2.4in). Time: Palaeocene–Oligocene. Order: Rotaliida.

3 **Cryptoprora** resembles a lacy, pointed hat with ribbons hanging from the rim. Size: 0.1–1mm (0.004–0.04in). Time: Eocene onward. Class: Actinopoda. Subclass: Radiolaria (protozoans with glassy, perforated skeletons shaped like hats, urns, spheres, etc).

4 **Tintinnopsis** has a bell-shaped skeleton with a pointed "tail". Size: 0.1–0.2mm (0.004–0.008in). Time: Recent. Phylum: Ciliata (ciliates), fringed by fine hairs (cilia) whose rhythmic beating helps them swim.

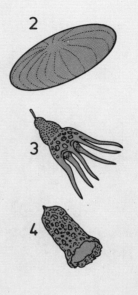

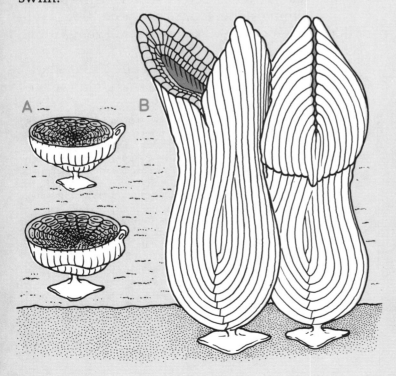

Problematic organisms
Some fossil organisms fit no known group of protists, plants, or animals. They include the tiny aquatic acritarchs and chitinozoans, also the larger, many-celled Petalonamae, shown here.
A Juvenile specimens
B Reconstructed group (smaller than life size) on a sandy sea floor, in late Precambrian times. Bodies comprised funnel-shaped colonies of branching tubes. The central cavity and scattered "needles" in the skeleton show that the Petalonamae had certain features found in sponges.

Parazoans

Sponges and the sponge-like archaeocyathids make up the animal subkingdom Parazoa. Parazoans are many-celled animals that lack true tissues or organs and can look misleadingly like plants. Sponges have simple bag-like bodies open at one end and anchored to the sea bed at the other. Mineral spicules (tiny rods) reinforce the body wall, which is pierced by tiny holes. The body cavity is lined with cells equipped with whips that draw in water through the holes. A sponge extracts food particles and oxygen from this water, and then "whips" drive the used water out through the main body opening, the osculum.

Some sponges live singly, others form colonies measuring from under 1 centimetre to over 1 metre across. Sponges grow as "vases", branches, or rock-encrusting blobs. Fossil skeletal remains suggest that sponges evolved from protozoans 700 million years ago. Some prehistoric sponges built reef-like sea-bed rocks. Scientists divide the phylum Porifera (sponges) into four classes according to type of skeleton. Our examples show a fossil genus from each class.

Archaeocyathids were usually cup-shaped like a fossil coral, but with a perforated wall like a sponge

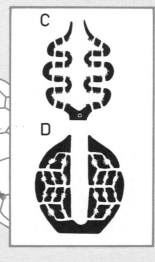

Sponge features
Shown here are different types of sponge, with features found in living specimens.
A Simple sponge: cutaway view of part of a colony, shown much enlarged
a Incurrent pores, letting water enter
b Osculum, letting water out
c Pore cell
d Spicule
e Collar cell
f Covering cell
B Simple sponge, actual size
C More advanced sponge, with a folded wall (shown in section).
D Complex sponge, with many canals and chambers (section).

(in fact most had a double wall). Experts put them in their own phylum, Archaeocyatha, with two classes. Archaeocyathids evidently lived on sea beds, much as sponges do, and flourished in warm, clear seas, worldwide – but only in Early–Mid Cambrian times.

1 **Siphonia,** about 1cm (0.4in) long, has persisted since Cretaceous times. Class: Demospongea (horny sponges largely reinforced by networks of horny fibres).

2 **Coeloptychium** was a mushroom-shaped sponge common in Late Cretaceous Europe. Diameter: 7.5cm (3in). Class: Hyalospongea (glass sponges with skeletons made up of six-rayed spicules).

3 **Polytholosia** resembled a string of pearls 5cm (2in) long. Time: Triassic. Class: Calcispongea (sponges with two- to four-rayed limy spicules).

4 **Chaetetopsis,** 4cm (1.6in) long, lived from Ordovician to Tertiary times. Class: Sclerospongea (coralline sponges with a skeleton of pin-shaped spicules).

5 **Tabellaecyathus,** was a cone-shaped archaeocyathid, 4cm (1.6in) long. Time: Cambrian. Class: Irregulares (cup- and disc-shaped forms, often irregularly shaped).

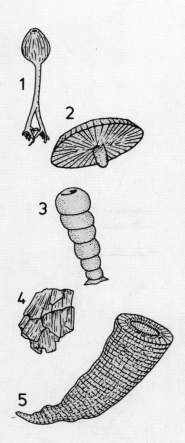

Spicules and fibres
Scattered spicules are often the only clues to the fossil sponges in a rock layer. Here we show spicules from two classes of sponge, and horny fibres from a third.
a Calcareous spicules belonging to the Calcispongea
b Siliceous spicules belonging to the Hyalospongea
c Horny fibres belonging to the Demospongea, and rare as fossils because they readily dissolve

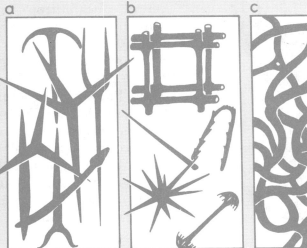

©DIAGRAM

Coelenterates

Coelenterates or "hollow gutted animals" are mostly jelly-like sea creatures. Each consists of many cells organized as tissues to produce a central body cavity with a mouth surrounded by tentacles and stinging cells for catching prey. Coelenterates are less primitive than sponges but have no central nervous system or systems for breathing, circulation, or excretion.

Cnidaria is the only phylum known from fossils. Cnidarians include jellyfishes, hydrozoans, sea anemones, and corals. Free-swimming cnidarians (medusas) produce polyps "rooted" to the sea floor. Polyps produce medusas, and so on. Cnidarians evolved about 700 million years ago. Limy cups enclosing colonial corals and hydrozoans in time built limestone reefs that formed rock layers.

Fossil cnidarians include these five examples. Numbers 3–5 are from the class Anthozoa (corals and sea anemones).

1 **Conularia** produced a polyp in a tall, hard, hollow "pyramid" inverted on the sea floor. Tentacles probably surrounded the broad mouth end Length: 6–10cm (2.4–4in). Time: Cambrian–Triassic. Class: Scyphozoa (scyphozoans) – jellyfishes etc.

2 **Millepora** comes in colonies of feeding polyps in central tubes, surrounded by protector polyps in other tubes. Size: 7cm (2¾in). Time: Recent. Class: Hydrozoa (hydrozoans).

3 **Charnia,** a sea pen, formed colonies of "feathers" growing from the sea bed. Height: 14cm (5½in). Time: Precambrian (about 700 million years ago). Subclass: Octocorallia (Alcyonaria), soft corals.

4 **Streptelasma** was a rugose ("wrinkled") coral, named from horizontal wrinkles on the skeleton's outer wall. Height: about 2.5cm (1in). Time Ordovician–Silurian. Subclass: Zoantharia (stony corals). Order: Pterocorallia (rugose corals), mostly solitary, now extinct.

5 Halysites, a "chain coral", produced chains of tubes stuck together at the sides. Size: 5cm (2in) across. Time: Ordovician–Silurian. Subclass: Zoantharia. Order: Tabulata (tabulate corals, named from internal horizontal plates, or tabulae), extinct since Permian times.

6 Acropora produces a light, branching skeleton about 30.5cm (1ft) long. Time: Eocene–Recent. Subclass: Zoantharia. Order: Cyclocorallia (Scleractinia), the major modern reef-builders, probably evolved from naked sea anemones.

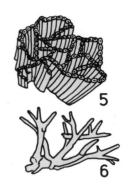

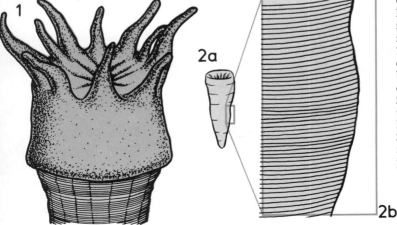

Scarce survivors
This fossil "jellyfish" came from Carboniferous rocks in Belgium. Fossils of soft-bodied beasts like this are rare. They survive only in rocks such as shales or limestones formed from fine-grained particles of mud.

Growth of coral
Here we show how coral polyps form cups that have survived as fossils.
1 A living coral polyp in its coral cup. The creature builds its cup by adding a daily layer of calcium carbonate to its top.
2a Growth bands plainly show up on this Silurian coral cup, real height less than 2.5cm (1 inch).
2b This magnified view shows a few of the daily growth rings that produced the cup.

53

Molluscs 1

Molluscs ("soft" animals) include snails, clams, squids, and other creatures with a shell and muscular foot, or parts derived from these, and a digestive tract. Most live in the sea: grazing on algae, filtering food particles from mud or water, or hunting. Molluscs probably evolved in Precambrian times from animals like flatworms. By Carboniferous times, some molluscs invaded land and fresh water. No group of animals except the arthropods have diversified more. The phylum Mollusca produced many thousands of species, grouped into two subphyla with nine known fossil classes. Our examples all lived in the sea.

1 **Chiton** (coat-of mail shell). Chitons resemble flattened wood lice guarded by a shell of seven or eight overlapping plates. They cling to rocks and graze on algae. Chitons date from Late Cambrian times. Only known fossil class: Polyplacophora.

2 **Scenella** was limpet like, with a cap-shaped single shell. Size: 1cm (½in) across. Time: Cambrian. Class: Monoplacophora (maybe ancestral to the squids, octopuses, and other cephalopods).

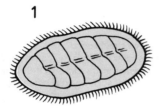

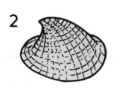

Four types of mollusc
Shown here are sections through four of the main types of mollusc described on pages 54–57.
A Chiton (coat-of-mail shell)
B Clam (bivalve mollusc)
C Snail (gastropod)
D Squid (cephalopod)
Despite differences, all share the same basic body plan, with the following common features.
a Shell (usually the only part preserved in fossils)
b Foot
c Digestive tract

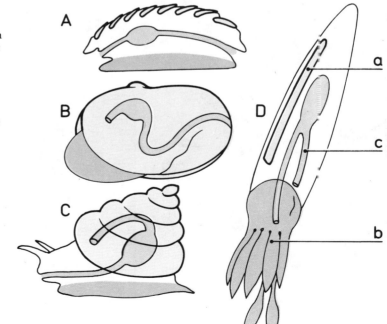

3 Pleurotomaria was a primitive snail with a spiral, pointed shell about 5cm (2in) high. Time: Jurassic–Cretaceous. Class: Gastropoda (molluscs with head, foot, and a mantle that secretes a bowl-shaped or spiral shell). They feed with a radula – a horny ribbon, armed with rows of rasping "teeth".

4 Dentalium lives in a slim, tusk-like shell and burrows head-first in soft sediment, extending thread-like tentacles to catch small organisms. Shell length: up to 12cm (4.7in). Class: Scaphopoda (tusk shells), known since Ordovician times.

5 Conocardium had a shell about 6cm (2.4in) across. Time: Ordovician–Permian. Class: Rostroconchia – extinct molluscs with a fused two-part shell. Rostroconchs perhaps gave rise to bivalves.

6 Mytilus, the common mussel, is a bivalve, with a hinged, two-part shell about 5cm (2in) long. Time: Triassic–Recent. Class: Lamellibranchia (bivalves) – headless molluscs with a hinged, two-valved shell.

3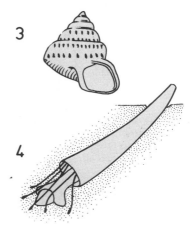

4

5

6

Unlikely bivalves
This restoration shows a sea-bed colony of rudists, 100 million years ago. In these bivalves one valve was a lid that opened to let in food and water. Rudist shells built reefs before these molluscs died out at the end of the Cretaceous Period.

©DIAGRAM

Molluscs 2

Squids, octopuses, and their kin form the cephalopods ("head footed"): the most highly developed class of molluscs. These big-brained, keen-eyed sea beasts can swim backward fast by squirting water forward. Tentacles at the head end seize prey and feed it to the beak-like jaws. Some cephalopods have an outside shell, some an inner shell, others none at all. Cephalopods range from a species only millimetres long to the giant squid – at up to 22 metres (72ft) long, the largest living invertebrate. Cephalopods probably evolved in Cambrian times, from gastropod-like molluscs. Our examples come from the four cephalopod subclasses: nautiloids, bactritoids, ammonoids, and coleoids. Fossils show that extinct ammonoids and (coleoid) belemnites were abundant in certain Mesozoic times. Besides cephalopods, there were two more mollusc classes: coniconchs and calyptoptomatids. These were small beasts with straight, cone-shaped shells. The last died out in Permian times.

A living nautiloid
Clues to fossil nautiloids come from the surviving *Nautilus*.
A Head-end, revealing the body in the outer chamber
B Side view, with shell cut open to reveal its structure
a Tentacles
b Jaws
c Gut
d Heart
e Gill
f Valve controlling water used for jet propulsion
g Siphuncle – tube connected to the inner chambers
h Chambers, filled with gas or air (*Nautilus* pumps water in or out to alter buoyancy, and so sink or rise)

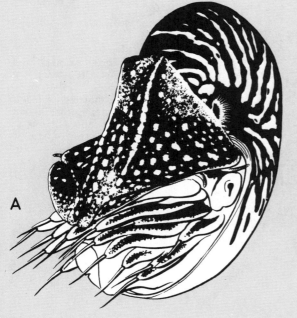

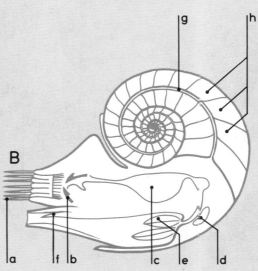

1 **Orthoceras** had a straight shell 15cm (6in) long.
Time: Ordovician–Triassic. Subclass: Nautiloidea
(nautiloids) – cephalopods with a straight or curved
shell divided into chambers. Nautiloids persisted
from Cambrian times to today.

2 **Cyrtobactrites** had a tusk-shaped shell. Length:
4.5cm (1.8in). Time: Early Devonian. Subclass:
Bactritoidea (ancestors of ammonoids).

3 **Anetoceras** had a loosely coiled shell about 4cm
(1.6in) across with zig-zag growth lines. Time: Early
Devonian. It belonged to the goniatites, early
members of the subclass Ammonoidea. Ammonoids
resembled nautiloids but had shells with wrinkled
growth lines between chambers.

4 **Stephanoceras** had a coiled, ribbed, disc-shaped
shell 20cm (8in) across. Time: Mid Jurassic. It
belonged to the ammonoids called ammonites.
Thousands of species teemed in Mesozoic seas.

5 **Gonioteuthis** was a belemnite, with a long,
squid-like body known only from the rear end's
fossilized, bullet-shaped internal guard. Guard
length: 7cm (2.8in). Time: Late Cretaceous. Order:
Belemnitida, in the subclass Coleoidea (squids,
cuttlefish, octopuses, and belemnites).

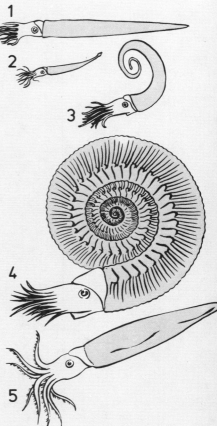

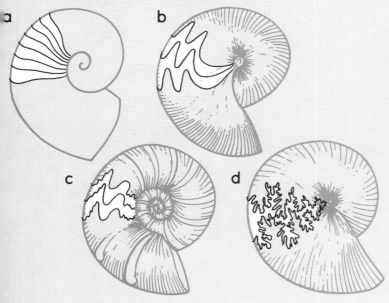

Shell sutures
Nautiloids, goniatites, ceratites
and ammonites respectively
evolved ever more complex
sutures – lines formed where
internal partitions met the
outside shell. These four cut-
open shells show stages in this
trend.
a *Nautilus* (simple sutures)
b *Goniatites* (zig-zag sutures)
c *Ceratites* (wavy sutures)
d *Phylloceras* (very complex
sutures)
Complex sutures might have
made shells strong enough to
endure water pressure deep
down. Many ammonites were
arguably bottom feeders.

©DIAGRAM

Worms

Soft-bodied beasts such as worms are seldom fossilized, but we know of prehistoric kinds from tracks, tunnels, tubes, jaw remains, and even impressions of their bodies left in fine-grained rocks. Such clues reveal that worms were crawling on and in, and swimming just above, the floors of shallow coastal waters some 700 million years ago, where South Australia's Ediacara Hills now stand Southwest Canada's famous Burgess Shales reveal that many sea worms had evolved 550 million years ago. Of our examples, number 1 comes from the phylum Aschelminthes (roundworms, built on simple lines), 2–5 come from the Annelida (worms divided into many segments, often with complicated jaws, and tentacles projecting from the head). Annelids probably gave rise to millipedes and other arthropods.

1 **Gordius**, a "hair worm", resembles a living thread up to 15cm (6in) long. Adults wriggle in ponds and ditches worldwide. Young live inside aquatic insect larvae; adults invade land insects, then return to water to breed. Time: Eocene on. Phylum: Aschelminthes. Class: Nematomorpha.

1

Durable mouth parts
Prehistoric polychaete worms are known mostly from their mouth parts – the only hard parts of their bodies, and so the only structures tending to be fossilized. Such fossils are called scolecodonts.
A Enlarged section through a polychaete's head (items **b–e** are the hard parts likely to be fossilized)
a Mouth
b Mandible
c Maxilla
d Dental plate
e "Pincer"
B Five Permian scolecodonts 15 times larger than life

A

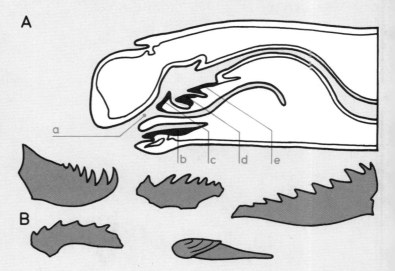

B

2 **Spriggina,** one of the first known worms, was probably an annelid. It was slim and flexible, with a long, strong, curved head shield, and up to 80 "limbs" with spiny ends. In Late Precambrian times it swam offshore in what is now Australia. Length: up to 4.5cm (1.8in).

3 **Dickinsonia** had a broad, flat, oval body crossed by 20 to 550 ridges. Length: 0.25–60cm (0.1in–2ft). It lived in a Late Precambrian sea in what is now Australia. It resembled some flatworms but was probably an annelid.

4 **Canadia** was a marine annelid with bristles, and long "legs" projecting from its sides. Time: Mid Cambrian. Place: south-west Canada. Class: Polychaeta (segmented worms with "legs" and bristles). Order: Errantia (mainly active polychaetes, with a well-defined head and jaws).

5 **Serpula** lives in a white, limy tube glued to an undersea shell or stone. Tentacles projecting from the head catch passing particles of food. Tube length: about 8cm (3in). Time: Silurian to present. Class: Polychaeta. Order: Sedentaria (jawless worms living in a tube or tunnel).

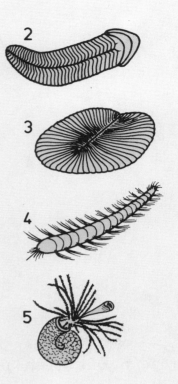

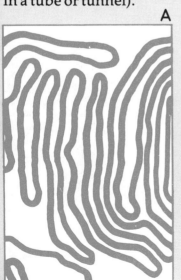
A

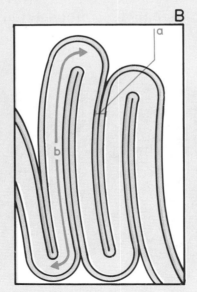
B

Worm trails
Some rocks preserve worm tracks made in soft mud that later hardened into stone.
A An unknown worm-like beast left this fossil track, seen from above. The creature ploughed through underwater sediments. Most likely it was feeding on organic particles or tiny organisms.
B This reconstruction shows how track **A** took shape.
a The width of churned up sediment fixed the distance of adjoining trails.
b The lengths of straight sections hint at the length of the worm that formed the U-turns seen in the track.

©DIAGRAM

59

Arthropods 1

Most known kinds of animals belong to the phylum Arthropoda: "jointed legged" animals. Arthropods include the extinct trilobites, as well as insects, spiders, crabs, centipedes, and relatives. All have jointed limbs, and an exoskeleton shed from time to time as they grow. Arthropods probably evolved from annelid seaworms. Early arthropods lived in the sea but some were colonizing land by 400 million years ago. These two pages give fossil examples of major groups based mostly on land. Onychophorans ("velvet worms") may be a primitive link with the arthropods' annelid worm ancestors. Many-legged myriapods (centipedes and millipedes) were among the first land arthropods. From early, wingless insects came the bees and butterflies – winged pollinators evolving with, and feeding on, the flowering plants. Most fossil insects date from the last 60 million years. Some of the best specimens occur in coal and amber.

1 **Aysheaia** looked like a worm with a pair of stubby legs on each body segment and a pair of "feelers". It was arguably related to the living velvet worms but lived below the sea, in Mid Cambrian times. Subphylum: probably Onychophora.

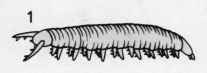

1

Arthropleura
At 1.8m (6ft) *Arthropleura* was the largest-ever land arthropod. This huge, flat "millipede" munched rotting vegetation on the floors of Carboniferous forests.

2 Latzelia has been called an early centipede. It had poisoned fangs, a flat back and a pair of walking legs on each of its many body segments. It roamed damp forest floors, hunting worms and soft insects. Time: Carboniferous. Subphylum: Myriapoda (centipedes, millipedes).

3 Rhyniella, the first known (wingless) insect, was a springtail, about 1cm (½in) long. Springtails live in soil, browse on decaying plants, and flip into the air if scared. Time: Devonian, about 370 million years ago. Subphylum: Insecta (insects) – small invertebrates with six legs and a three-part body (head, thorax, abdomen). Many undergo great body changes (metamorphosis) as they grow.

4 Meganeura, the largest-known winged insect, had a wing span of up to 70cm (27½in). Time: Late Carboniferous. *Meganeura* belonged to the Megasecoptera, primitive winged insects unable to fold back wings held at rest.

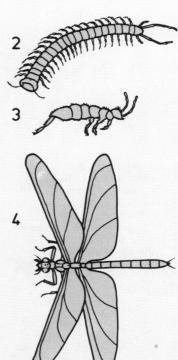

2

3

4

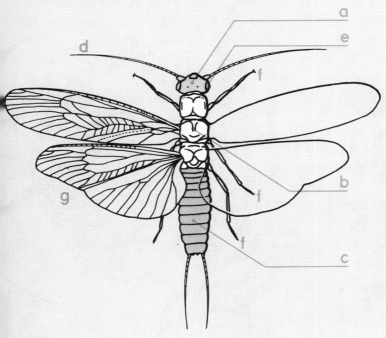

Insect features
This stonefly illustrates typical insect features, such as the three-part body. Early flying insects had two pairs of wings of equal size. They could flap each pair separately, but only up and down; they could not fold their wings back at rest. In such later insects as butterflies, both pairs are coupled. In flies, hind wings are tiny and only serve as balancers.
a Head
b Thorax
c Abdomen
d One pair of antennae
e Compound eyes
f Three pairs of legs
g Wings reinforced by "veins"

Arthropods 2

Trilobites ("three lobed") were marine arthropods resembling wood lice. Their name comes from a central raised ridge along the back, flanked by flattish side lobes. There was a head shield and an armoured thorax and tail, both divided into many segments. Each of these sprouted a pair of limbs designed for walking, swimming, breathing, and handling food. Trilobites crawled or swam, and many could curl up if threatened. Scientists know of several thousand genera and perhaps 10,000 species, from less than 4mm (0.1in) to 70cm (28in) long. Trilobites lived from about 600 to 250 million years ago. They dominated shallow seas in the Cambrian Period. Ordovician genera included many specialized species with bizarre spines or knobs. These pages show a few contrasting kinds.

Trilobite body plan
Calymene illustrates the basic body plan of trilobites.
A Cephalon, a head with a central hump
B Thorax, consisting of a number of jointed segments
C Pygidium, a tail of fused segments articulated with the thorax

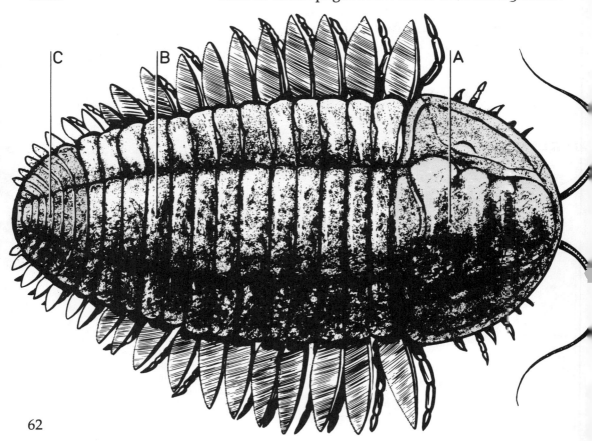

1 **Ampyx** was lightly built and filter fed. Length: 4cm (1.6in). Time: Ordovician.
2 **Encrinurus** had a relatively large, heavy head with eyes on stalks, and was a bottom dweller. Length: 5cm (2in). Time: Ordovician-Silurian.
3 **Lonchodomas** had a long spike jutting forward from the head. Length: about 5cm (2in). Time: Ordovician.
4 **Pliomera** Tooth-like ridges jutting from the front of the head interlocked with the tail tip when this animal rolled up. Length: about 4cm (1.6in). Time: Ordovician.
5 **Trimerus** was elongate and lived by burrowing in mud. Length: 8cm (3in). Time: Silurian-Devonian.

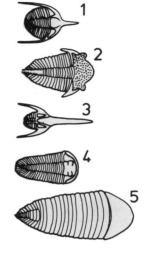

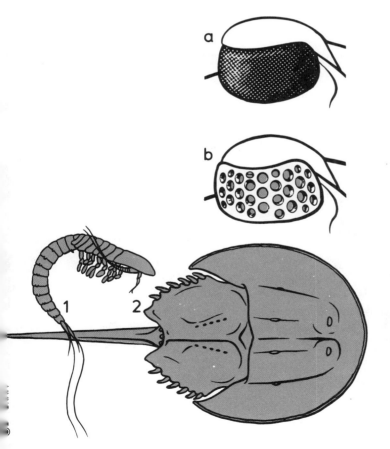

Amazing eyes
Trilobites were among the first known animals with efficient eyes. These had many calcite-crystal lenses fixed at different angles to register movement and light from different directions.
a Compound or holochroal eyes consisted of 100 to 15,000 closely packed hexagonal lenses and resembled insects' eyes.
b Schizochroal eyes featured groups of lenses, relatively few in number.
Living fossils
Trilobites died out more than 230 million years ago, but we show two kinds of living arthropods which have been regarded as close kin of these creatures.
1 Cephalocarids are tiny, primitive, shrimp-like animals with segmented bodies much like trilobites'. First found in 1955, cephalocarids may be their nearest living relatives.
2 The horseshoe crab looks a bit like a trilobite and has a larval stage reminiscent of a young trilobite. This type of "crab" has lived for over 300 million years.

Arthropods 3

Spiders, scorpions, horseshoe crabs, and the extinct sea scorpions make up the subphylum Chelicerata. Prehistoric chelicerates included arthropods longer than a man – fish-eating terrors of ancient seas.

Chelicerates get their name from the two chelae ("biting claws") in front of the mouth. Behind these come a pair of pedipalps ("foot feelers"), used by horseshoe crabs as legs, by scorpions for seizing prey, by male spiders to grip when mating. Then come four pairs of legs. The two-part body's limbless abdomen often ends in a flat or spiky tail. Horseshoe crabs and sea scorpions had gills and breathed under water. Modern spiders and scorpions breathe atmospheric air with help from "lung-books" or air holes called tracheae.

Chelicerates appeared in the sea over 560 million years ago. They gave rise to sea scorpions soon after 500 million years ago. True scorpions evolved about 440 million years ago. The first known spiders were catching insects 370 million years ago. Most chelicerates grow a rather soft body covering, so few survive as fossils.

Spider in amber
Above: sticky sap trapped a spider 30 million years ago. Below: the sap hardened into amber which still preserves the spider's life-like body.

Eurypterid body plan (right)
Shown here are two life-size views of the Silurian "sea scorpion" *Eurypterus*.
A View of underside:
a Chelicerae (appendages with jointed pincers – long and formidable in beasts like *Pterygotus* which was capable of catching fishes)
b Four pairs of walking legs
c Large "paddles"
d Genital appendages
B View of upper side:
a Prosoma (forepart)
b Opisthosoma (hind part, comprising 12 articulated segments) including:
c Telson ("tail")

1 **Palaeolimulus** had a broad, horseshoe-shaped forepart with two large, many-faceted eyes. The narrow abdomen ended in a spiky "tail". Length: 6cm (2.4in). Time: Permian. Place: shallow seas. Class: Merostomata. Subclass: Xiphosura (horseshoe crabs).

2 **Pterygotus,** one of the largest-ever arthropods, was a formidable predator. Its forepart had two long, strong pincers, eight legs, and two large, broad paddles for swimming. Twelve segments and a tail formed the long hind end. Length: up to 2.3m (7ft 4in). Time: Silurian. Place: seas. Class: Merostomata. Subclass: Eurypterida (eurypterids or sea scorpions).

3 **Palaeophonus,** a late Silurian scorpion, had big pincers and a large sting on its tail. It might have been the first land animal. Class: Arachnida (spiders and scorpions). Order: Scorpionida (scorpions).

4 **Arthrolycosa** was a large, long-limbed early spider. It had eight legs and eight eyes and probably attacked insects with help from poisoned chelae ("fangs"). Time: Carboniferous. Class: Arachnida. Order: Araneae (spiders).

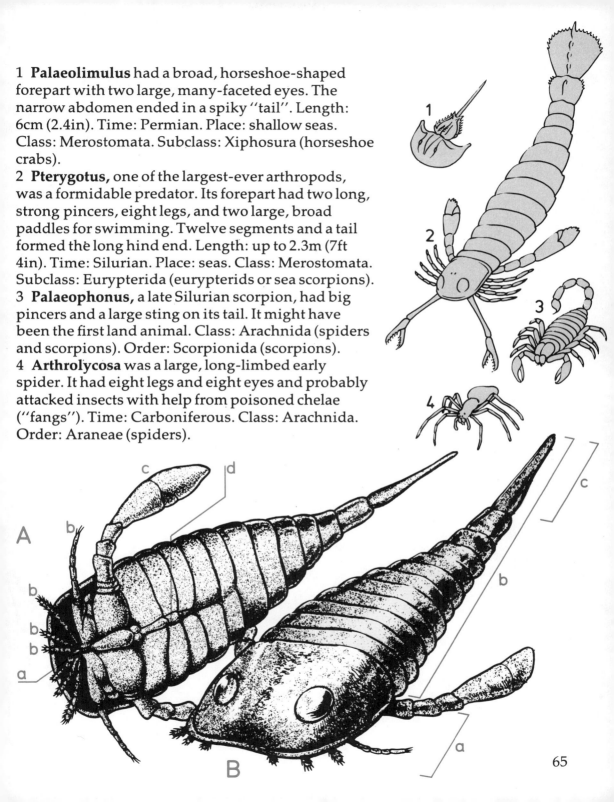

65

Arthropods 4

Crustaceans include crabs, barnacles, and relatives with flexible shells. Members of their subphylum, the Crustacea, have three main body parts (the first two often look like one), two pairs of antennae, and many more appendages. The limbs are forked. Crustaceans evolved by 650 million years ago. From early forms arose nine classes, all mostly found in water. Three produced a rich variety of fossils. These classes were the tiny ostracods, the cirripedes (barnacles and kin), and the malacostracans (including lobsters, crabs, and shrimps). Lobster-like crustaceans date from Triassic times and gave rise to crabs in the Jurassic Period.

1 **Cypridea** was a freshwater crustacean resembling a microscopic clam. Its body lay inside a two-valved shell with straight top and bottom edges and a knobbly surface. Time: Mid Jurassic–Early Cretaceous. Class: Ostracoda – ostracods (mostly marine organisms, valuable as guide fossils from Cambrian times onward).

Ostracod body plan
Experts identify fossil ostracods from the shapes and patterns of their shells. In life these small crustaceans (none larger than a bean) looked like the living specimen seen here enlarged. Such animals are found in seas and ponds.
a Hinged shell, shown in section; when shut it protected all parts of the body
b Short body
c Antennae (used as limbs)
d Mandibles (mouthparts)
e Maxillae (mouthparts)
f Trunk appendages
g Furca ("tail")

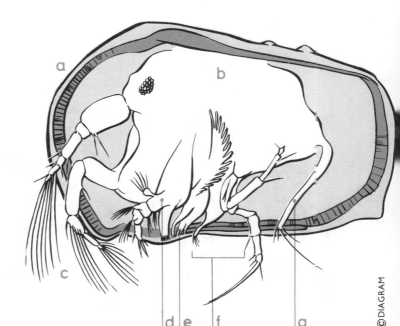

©DIAGRAM

2 Balanus, an acorn barnacle, lives inside a dome-like shell of six white limy plates, built on a tidal rock. At high tide curved, feathery appendages move in and out between shell plates to pull in scraps of food, and water. Size: about 1cm (½in) across. Time: Oligocene onward. Class: Cirripedia (barnacles) – crustaceans that fix their heads to rocks or other solid objects.

3 Aeger was a long-tailed ten-legged crustacean with a long bill-like rostrum and long antennae. Length: about 12cm (4.7in) excluding rostrum. Time: Late Triassic–Late Jurassic. Class: Malacostraca (advanced crustaceans with stalked eyes and a "shell" usually covering the head and thorax). Order: Decapoda (ten-legged crustaceans, including shrimps, crabs, and lobsters).

4 Eryon had a big, broad, flattened, crab-like cephalothorax (fused head and thorax) but a longish abdomen. Crabs evolved from beasts like these that tucked the abdomen beneath the forepart of the body. Length: 10cm (4in). Time: Mid Jurassic–Early Cretaceous. Order: Decapoda.

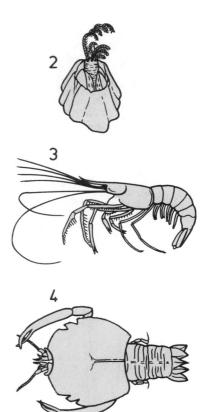

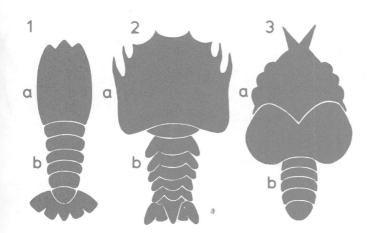

Crabs in the making
Shown here are the bodies of three prehistoric crustaceans (without their limbs).
1 *Eryma*
a Narrow cephalothorax (head and thorax)
b Abdomen as long as the cephalothorax
2 *Eryon*
a Broad, flat cephalothorax
b Long abdomen
3 *Palaeinachus*
a Big, broad cephalothorax
b Short, narrow abdomen

Tentaculates

This name is used for four great groups of lowly sea beasts. Tentaculates have a U-shaped gut and a mouth surrounded by a ring of tentacles, used for pulling tiny particles of food and water into the mouth. Most live rooted to the sea bed. Phoronids are worm-like creatures living in a hard protective tube. Bryozoans are tiny, soft-bodied beasts in hard limy or horny cases. Colonies form mats or miniature "trees" on underwater rocks or shells. Brachiopods ("arm footed" animals) resemble bivalve molluscs sprouting from a fleshy stalk, but shell valves cover the body from above and below, not from side to side. Conodonts ("cone teeth") are brown or greyish tooth-like microfossils from eel-like creatures.

Conodonts flourished from Cambrian to Triassic times. Brachiopods date from the Early Cambrian, bryozoans from the Ordovician, and phoronids from the Cambrian Period. All groups but conodonts have living representatives.

Conodont types (right)
Conodonts are preserved hard parts of soft-bodied animals resembling elvers (young eels), only recently discovered. These may not have been tentaculates at all. Their tiny tooth- and saw-like fossils came in several types. Three kinds appear here, much enlarged.
A Compound conodont with many "teeth"
B Simple conodonts
C Platform conodont

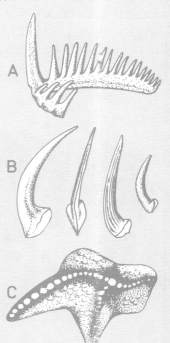

Evolved from worms (left)
Lingula's long worm-like body hints at the likely origin of brachiopods. Often called lampshells, they probably evolved from a tube-shaped worm that grew a pair of flat, protective shells.

1 **Phoronis** is a worm-like tube dweller about 20cm (8in) long. It lives buried in sand offshore. Tiny whip-like cilia on its tentacles drive water and food into its mouth. Time: maybe Cretaceous onward. Phylum: Phoronida (phoronids).

2 **Multisparsa** was a bryozoan forming tree-like colonies of tubes with touching sides. Colony size: 2cm (0.8in) across. Time: Mid Jurassic. Phylum: Bryozoa (sea mats).

3 **Lingula** lives inside a long, tongue-shaped shell growing from a strong stalk. It inhabits a burrow in soft undersea sediment. Diameter: 10cm (4in). Time: Ordovician onward (it is about the oldest living fossil). Phylum: Brachiopoda. Class: Inarticulata (brachiopods whose shell lacks a hinge).

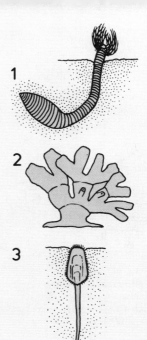

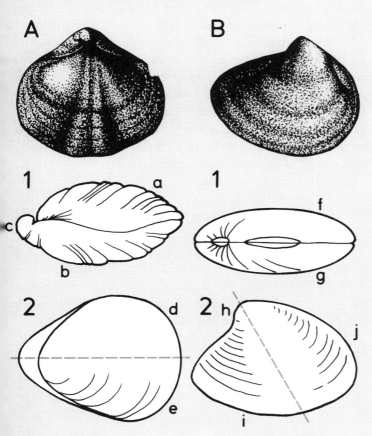

Brachiopods and bivalves
At first glance you might mistake a fossil brachiopod for a bivalve mollusc, another creature with a two-valved shell. These illustrations show how to tell both apart.
A Brachiopod: both valves seen edgewise (**1**), and one valve seen from above (**2**)
a,b Shell valves differ in size and curvature
c Pointed beak at hinge end
d,e Each half of one valve is a mirror image of the other
B Bivalve: both valves seen edgewise (**1**) and one seen from above (**2**)
f,g Shell valves alike in size and curvature
h No beak at hinge
i,j Valves not symmetrical

©DIAGRAM

Echinoderms and branchiotremes

Echinoderms ("spiny skinned" animals) include starfishes, sea urchins, and their relatives. These tend to have a five-rayed body with a skeleton of chalky plates just below the skin. Some sprout protective spines. Water flows through tubes inside the body, and pumps up many tiny tube feet tipped with suckers and used for walking, gripping, or as breathing aids. Echinoderms lack a normal "head", but have a well-developed nervous system. They probably gave rise to backboned animals by way of branchiotremes.

These small, soft-bodied, worm-like creatures with internal "gill baskets" include the pterobranchs and long-extinct but once abundant graptolites. Branchiotremes produced the beginnings of a notochord or "backstring" – the precursor of a backbone. Some experts rank them as a phylum, others consider them a subphylum of the Chordata, which includes all backboned animals. Echinoderms and many branchiotremes live on the sea bed. Branchiotremes use tiny tentacles to catch passing scraps of food. Echinoderms include hunters, grazers, scavengers, and filter feeders. Both phyla go back to Cambrian times. Examples 1–5 represent echinoderm subphyla. Examples 6–7 represent two classes of branchiotremes.

1 **Dendrocystites** looked like a double-ended thorn and probably laid flat on the sea bed, eating tiny organisms. Length: 9cm (3.5in). Time: Ordovician. Subphylum: Homalozoa (carpoids), extinct.

2 **Pleurocystites** sprouted two short, unbranched arms from a bud-shaped "cup" on a long stalk. Stalk height: 2cm (0.8in). Time: Ordovician. Subphylum: Blastozoa (blastoids and cystoids), extinct.

3 **Botryocrinus** was a flower-like sea-bed dweller with a stalk, crown, and feathery arms. Height: 15cm (6in). Time: Silurian. Subphylum: Crinoidea ("sea lilies"), Cambrian onward.

How tube feet work (above) This shows a section along a starfish arm. Water from a pipe inside fills bottles. Muscles squeeze them, driving water down into tube feet. The feet lengthen. As water flows back, each foot tip forms a sucker.

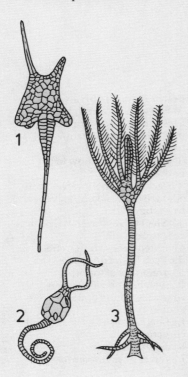

4 **Crateraster** had a broad central disc and five slim arms. Size: 10cm (4in) across. Time: Cretaceous. Subphylum: Asteroidea (starfishes and brittle stars), dating from Ordovician times.

5 **Bothriocidaris** resembled a small, prickly pincushion. Size: about 2cm (0.8in) across including spines. Time: Ordovician. Subphylum: Echinozoa (sea urchins and sea cucumbers), Ordovician on.

6 **Rhabdopleura** is a soft-bodied worm-like beast living in colonies in tubes on the sea bed. Size: 2–3mm (0.08–0.12in). Phylum: Branchiotremata (branchiotremes). Class: Pterobranchia, dating from Ordovician times.

7 **Dichograptus** was an eight-branched graptolite – a colony of tiny worm-like creatures, known from their flattened, "saw-edged" fossil tubes. Colony diameter: 6cm (2.4in). Time: Ordovician. Phylum: Branchiotremata. Class: Graptolithina (graptolites), extinct, dating from Cambrian to Carboniferous.

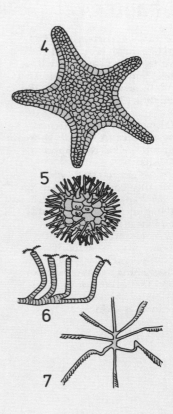

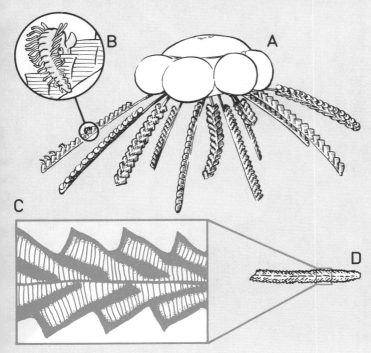

About graptolites
A Single-branched graptolites called diplograptids formed a colony buoyed up by a central float. Such colonies drifted at the surface of Early Palaeozoic seas. (Some others sprouted from the sea floor.)
B Each individual of a colony lived inside a cup-like theca.
C Much magnified view of fossil *Diplograptus* thecae. Infrared light shows details of structure poorly visible in ordinary light.
D *Diplograptus* colony, shown actual size.

71

Chapter 4

FOSSIL FISHES

With fishes we reach the first of five chapters on fossil vertebrates, or animals with a backbone. Vertebrates are just one of several subphyla in the phylum Chordata (chordates). But apart from hemichordates (also called branchiotremes), vertebrates are the only chordate group that has left important fossils.

This chapter starts with primitive jawless fishes in the class Agnatha. We briefly review the spiny fishes (Acanthodii) and the placoderms (Placodermi), both extinct. Lastly, we look at fossil examples of two classes flourishing today: the sharks and shark-like fishes (Chondrichthyes), and bony fishes (Osteichthyes).

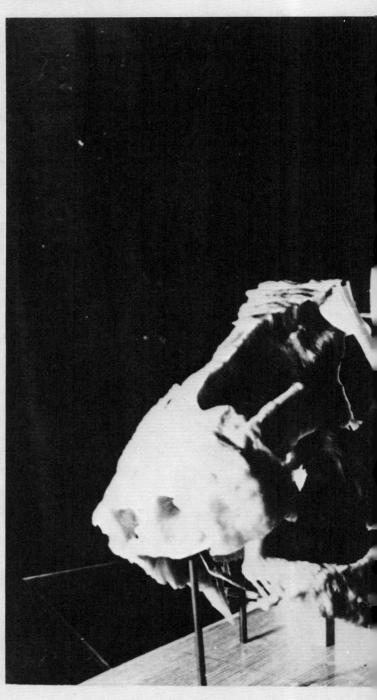

Swedish palaeontologist Erik Stensiö builds a three-dimensional model of a prehistoric fish's brain. In the 1920s, Stensiö's techniques of dissection and reconstruction of fossil brains revolutionized our understanding of some early fishes. (The Mansell Collection.)

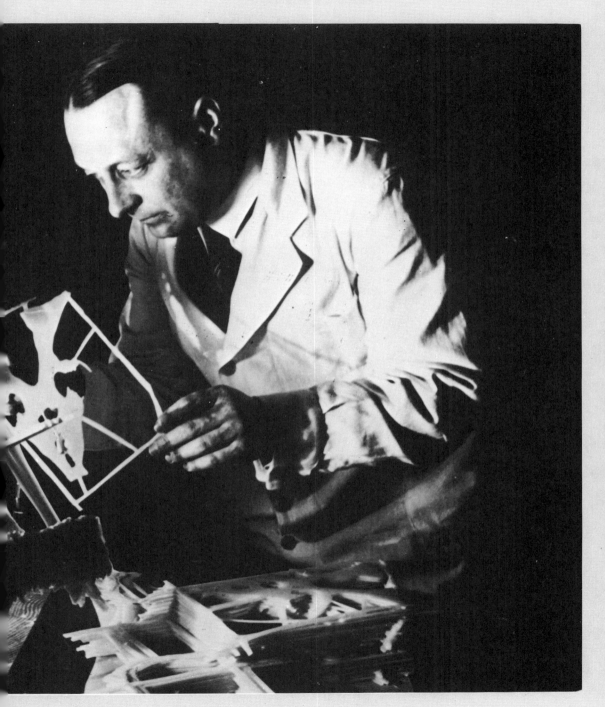

About fishes

Fishes are cold-blooded, backboned animals that live in water, breathing oxygen through gills. To swim forward most thrust water backward by waggling the tail and body.

Defined strictly, fishes have jaws, paired fins, and median fins (fins along the midline of the body). But this chapter includes creatures without jaws or true fins. Appearing over 500 million years ago, they were the first known backboned animals or vertebrates – creatures with a brain housed in a skull, and vertebrae (bones making up a backbone). Vertebrae provide bodily support and a protective channel for the spinal cord, which contains nerves connecting the brain with other parts of the body. From jawless fishes or close relatives there evolved four other

©DIAGRAM

Fish ancestor?
Creatures like this tiny living lancelet might have been the immediate ancestor of fishes. (A fossil lancelet 550 million years old is known from Canada.) Lancelets lack head, jaws, vertebrae, and paired fins, but have these fish-like features:
a Gills
b Notochord (flexible rod foreshadowing the backbone)
c Nerve cord
d Tail fin
(All animals with a notochord or backbone are chordates: members of the phylum Chordata. Lancelets belong to the subphylum Cephalochordata.)

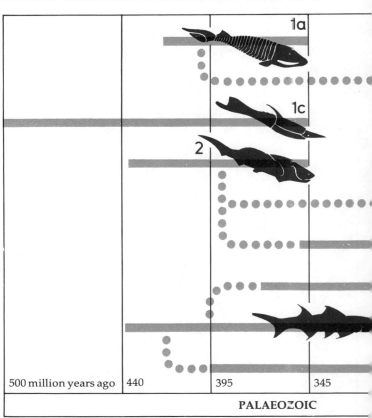

500 million years ago	440	395	345

PALAEOZOIC

major groups of fishes, two now extinct. Bony fishes, the most successful group, evolved a more flexible and mobile body than the fishes they replaced.

Fishes probably originated in the sea indirectly from echinoderms (see pp. 70–71), but most early kinds inhabited fresh water. Prehistoric species ranged from creatures a few centimetres long to a 14-metre-long (46ft) relative of the man-eating great white shark alive today.

Fossil fishes have been found in every continent. Appropriate rocks yield isolated bony plates, spines, scales, vertebrae, and teeth. But whole skeletons are rare, and many early fishes had gristly skeletons that seldom formed good fossils.

Family tree of fishes
In this family tree of all fishes, numbers show classes and letters show subclasses.
1 Agnatha (jawless fishes)
a Monorhina (with a nostril between the eyes)
b Cyclostomata (living jawless fishes)
c Diplorhina (no dorsal nostril)
2 Placodermi (placoderms)
3 Chondrichthyes (fishes with a gristly skeleton)
d Holocephali (shark-like fishes)
e Elasmobranchii (sharks)
4 Acanthodii (spiny fishes)
5 Osteichthyes (bony fishes)
f Actinopterygii (ray-fins)
g Sarcopterygii (fleshy-finned fishes)

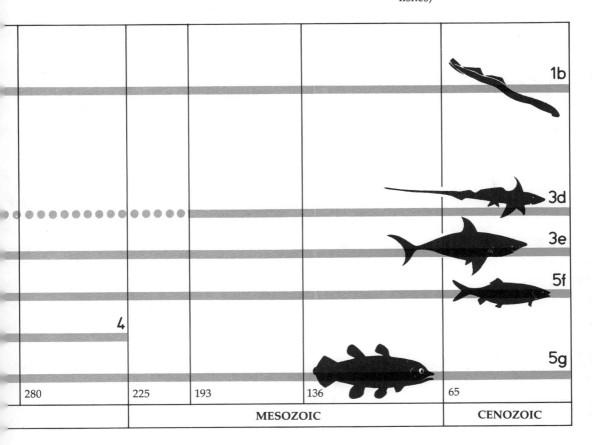

280	225	193	136	65
		MESOZOIC		CENOZOIC

Jawless fishes

Agnathans ("jawless fishes") included the first backboned animals. Body armour earns most extinct agnathans the collective name ostracoderms ("shell skins"). Armour probably saved some from being eaten by eurypterids (pp. 64–65). Ostracoderms flourished mainly in rivers and lakes about 510–350 million years ago. Most were small and lacked paired fins. Their armoured heads had jawless slits or holes for mouths, through which they sucked in water containing particles of food. Some scavenged in mud, others guzzled tiny organisms at the surface.

Ostracoderm fossils are plentiful in Late Silurian and Early Devonian rocks of Europe and North America.

Here is one example from each ostracoderm order.
1 **Hemicyclaspis** was an armoured fish with a solid bony head shield, usually backswept "horns", sensory fields in the head, eyes on top of the head and mouth below, body plated, triangular in cross section, tapering to the uptilted tail, with flat belly, one nostril, and many openings for gills. Length:

1

Cephalaspid body plan
These features include hints of the bottom-dwelling life of *Hemicyclaspis* and its relatives.
a Solid bony shield guarding head
b Eyes on top of head
c Mouth below head for sucking food from mud
d Sensory fields – aids for detecting movement

e Armour plates allowing side-to-side body movements
f Uptilted tail keeping head down
g Lateral fins helping balance
h Median fin stopping body rolling
i Flat belly

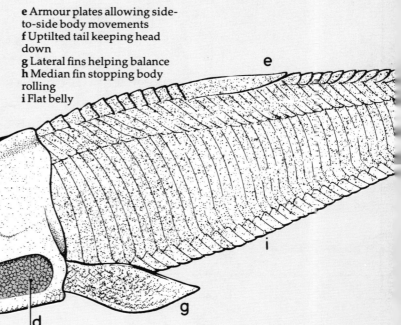

13cm (5in). Time: Late Silurian–Devonian. Place: northern continents. Order: Cephalaspida.

2 **Birkenia** was a small, deep-bodied ostracoderm, with small bony plates on the head, eyes at sides of the head, spines on the back, and tail angled downward. It had a fin along each side. Length: 10cm (4in). Time: Mid Silurian–Early Devonian. Place: Europe. Order: Anaspida.

3 **Pteraspis** was a fish with a long, rounded head shield, no visible nostril, long sharp "beak", slit-shaped mouth below, one gill opening per side, long spine on the back, no paired fins, and a downward-angled tail. Length 5.7cm (2¼in). Time: Early Devonian. Place: northern continents. Order: Pteraspida.

4 **Thelodus** was a flat fish covered with small, tooth-like denticles not flat plates. It had a long upper tail lobe and eyes on the sides of the head. Length: 18cm (7in). Time: Mid Silurian–Early Devonian. Place: Europe, North America. Order: Coelolepida.

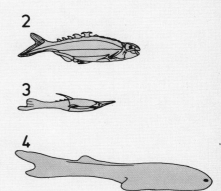

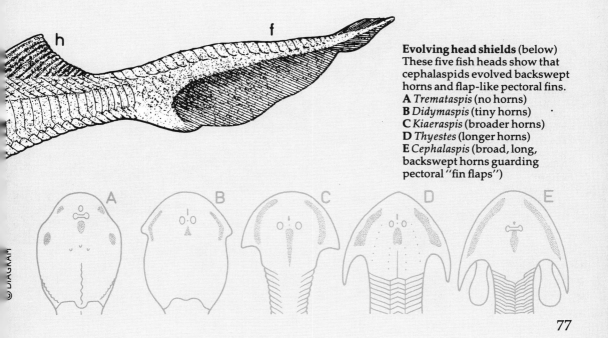

Evolving head shields (below)
These five fish heads show that cephalaspids evolved backswept horns and flap-like pectoral fins.
A *Tremataspis* (no horns)
B *Didymaspis* (tiny horns)
C *Kiaeraspis* (broader horns)
D *Thyestes* (longer horns)
E *Cephalaspis* (broad, long, backswept horns guarding pectoral "fin flaps")

Spiny fishes

Nicknamed "spiny sharks" or "spiny fishes", the acanthodians had stout spines along the leading edges of their fins but were neither sharks, nor bony fishes of a modern kind. These small freshwater species lived about 400–230 million years ago, and were the first known vertebrates with jaws. They had a blunt head, small, stud-like body scales, and a long, tapered upper tail lobe. Acanthodians swam at mid and surface levels. Some probably ate small freshwater invertebrates. Others preyed on jawless fishes.

Fossils crop up in Late Silurian to Permian rocks, and occur in almost every continent. Most are crushed flat in shale slabs. Early fossils are just spines and scales.

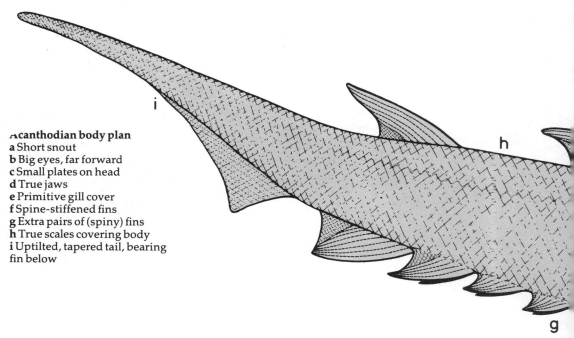

Acanthodian body plan
a Short snout
b Big eyes, far forward
c Small plates on head
d True jaws
e Primitive gill cover
f Spine-stiffened fins
g Extra pairs of (spiny) fins
h True scales covering body
i Uptilted, tapered tail, bearing fin below

Below are species from the three acanthodian orders: Climatiformes (primitive types); Ischnacanthiformes (types with reduced spines); Acanthodiformes (the last group, some degenerate).

1 Climatius had a short, deep body, and five pairs of extra fins below its belly. Length: 7.6cm (3in). Time: Late Silurian–Early Devonian. Place: northern continents. Order: Climatiformes.

2 Ischnacanthus was more advanced than *Climatius*, with fewer and slimmer but relatively longer and more deeply embedded spines. Time: Early–Mid Devonian. Place: Europe. Order: Ischnacanthiformes.

3 Acanthodes looked more eel-like than earlier acanthodians: it was partly scaleless, with fewer fins and spines and no teeth. Length: 30.5cm (1ft). Time: Late Devonian–Early Permian. Place: northern continents and Australia. Order: Acanthodiformes.

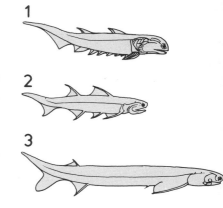

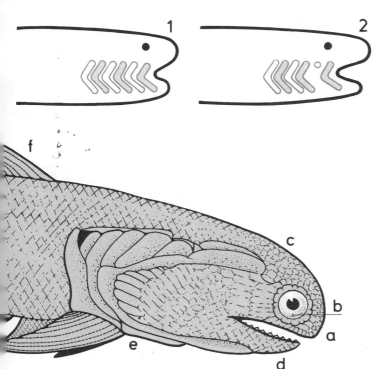

Jaws evolving (above)
Three diagrams show stages in the evolution of a fish's jaws, enabling it to open or close its mouth.
1 Jawless fish: bony gill supports (shown tinted) alternate with gills.
2 The first pair of gills shrinks to form a spiracle, a tiny hole for drawing in mud-free water.
3 The first pair of gill supports can now expand, and evolve into jaws.

Placoderms

The placoderms or "plated skins" were among the first fishes with jaws and paired fins. Bony armour covered the head and forepart of the body. In many, a movable joint between head and body armour let the head rock back to open the mouth wide The primitive jaws had jagged bony edges that served as teeth. The tail end usually lacked protection, even scales. Placoderms mostly swam with eel-like movements. Many lived on the sea bed. Some were the largest, most formidable creatures of their day. The group appeared in Silurian times, dominated Devonian seas (395–345 million years ago), and then died out under competition from sharks and bony fishes.

Placoderms are divided into seven orders. Here are examples of six of them, all Devonian.

1 **Gemuendina** was flat and broad. Length: 23cm (9in). Place: Central Europe. Order: Rhenanida (placoderms resembling skates).

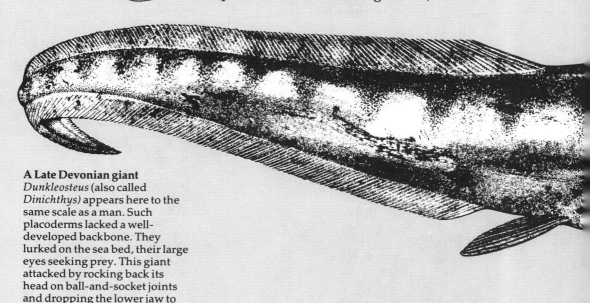

A Late Devonian giant
Dunkleosteus (also called *Dinichthys*) appears here to the same scale as a man. Such placoderms lacked a well-developed backbone. They lurked on the sea bed, their large eyes seeking prey. This giant attacked by rocking back its head on ball-and-socket joints and dropping the lower jaw to expose the bony cutting edges serving as its teeth.

2 **Lunaspis** had armoured skin all over and curved, bony shoulder spines. Length: 27cm (10½in). Place: Europe. Order: Petalichthyida.

3 **Dunkleosteus** (*Dinichthys*), a huge predator, could kill large fishes. Length: up to 9m (30ft). Place: North America and Europe. Order: Arthrodira (armoured placoderms with jointed necks).

4 **Rhamphodopsis** had grinding jaw plates and a big shoulder spine. Length: 10cm (4in). Place: Europe. Order: Ptyctodontida (small, armoured fishes).

5 **Phyllolepis** was flat, with little head armour. Length: 12.7cm (5in). Place: Australia, Europe, North America. Order: Phyllolepida (mostly flat and heavily plated placoderms).

6 **Bothriolepis** had a weak mouth, eyes on top of the head, and crab-like arms encasing the front fins. Length: up to 30cm (1ft). Place: found in most continents. Order: Antiarchi (small fishes with jointed, movable, spiny front fins).

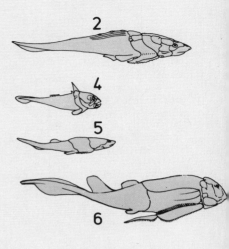

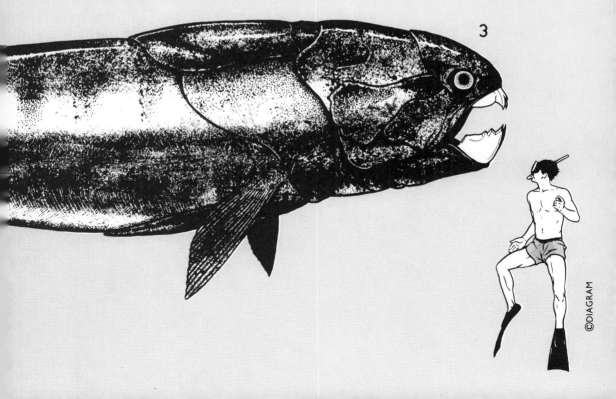

©DIAGRAM

Sharks and their kin

Sharks and shark-like fishes make up the Chondrichthyes – one of two classes of so-called higher fishes. Chondrichthyans have skeletons of tough, gristly cartilage, not bone, and tiny tooth-like scales. They have paired fins but lack gill covers, or swim bladders to adjust buoyancy.

Some are ferocious, streamlined killers with razor-sharp teeth. But skates, rays, and rat fishes include bottom-dwellers with low, broad teeth for crunching shellfish. Most kinds live in the sea.

Sharks probably evolved from placoderms 390 million years ago. They evolved fast, but many kinds died out. Shark fossils occur worldwide. Many are just teeth, fin spines, or tooth-like denticles from skin. The soft skeletons have mostly rotted. But fine-grained Late Devonian Cleveland shales preserve fine specimens of early sharks along Lake Erie's southern shore.

These fishes represent five chondrichthyan orders.
1 **Cladoselache** had a torpedo-shaped body, short snout, big eyes, broad-based fins, and long upper tail lobe. Some individuals had a spine on the back. Length: 50cm–1.2m (1ft 8in–4ft). Time: Late Devonian. Place: Europe and North America. Order: Cladoselachiformes (extinct ancestral sharks).

Early and modern
Here we compare some primitive features of *Cladoselache* with (in parentheses) those of fully modern sharks.
a Jaws at front of head (jaws on underside of head)
b Upper jaw fixed to braincase at back and front (upper jaw fixed to braincase at back only, allowing mouth to gape wide)
c Snout short and rounded (head pointed)
d Broad-based triangular fins (more mobile fins with narrow bases)
e Horizontal fin "rudders" near tail (no such fins)
f Torpedo-like body (similar)
g No claspers on pelvic fins (claspers on males' pelvic fins grip females during mating)

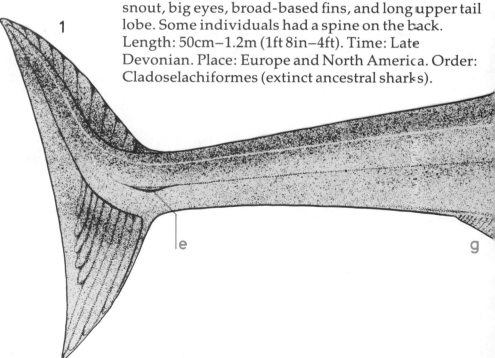

2 **Xenacanthus** had a long dorsal fin, long tail ending in a point, and a spine jutting back from its head. Length: 76cm (2ft 6in). Time: Late Devonian–Mid Permian. Place: Americas, Europe, and Australia. Order: Pleuracanthiformes (early freshwater sharks).

3 **Hybodus** had narrow-based, manoeuvrable fins, and a small anal fin. Length: over 2m (6ft 6in). Time: Late Permian–Early Cretaceous. Place: worldwide. Order: Selachii (modern sharks and close kin).

4 **Aellopos** was a flat fish with wing-like fins and whiplash tail. Length: 1.5m (5ft). Time: Late Jurassic. Place: Europe. Order: Batoidea (skates and rays).

5 **Ischyodus** had a stout dorsal spine, wing-like pectoral fins, and whip-like tail. Length: 1.5m (5ft). Time: Mid Jurassic–Palaeocene. Place: worldwide. Order: Chimaeriformes ("rat fishes").

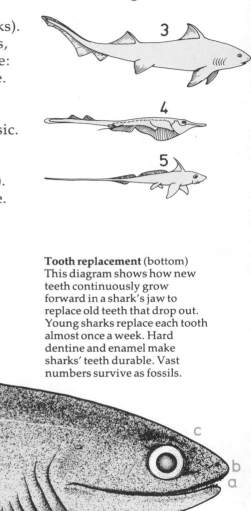

Tooth replacement (bottom) This diagram shows how new teeth continuously grow forward in a shark's jaw to replace old teeth that drop out. Young sharks replace each tooth almost once a week. Hard dentine and enamel make sharks' teeth durable. Vast numbers survive as fossils.

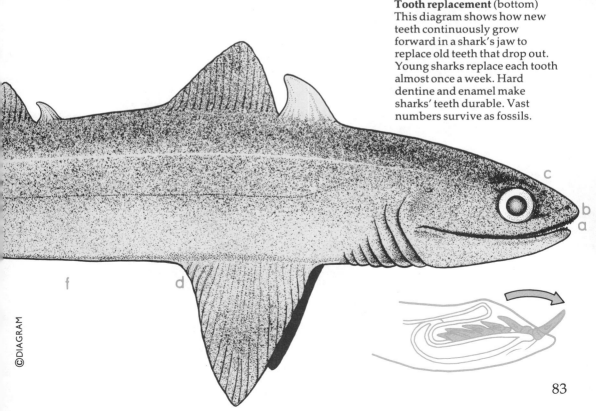

©DIAGRAM

83

Bony fishes 1

Ray-finned fishes – actinopterygians – account for almost all fishes now alive. Their ancestors grew far more plentiful and varied than the fleshy-finned fishes, the other subclass of Osteichthyes or bony fishes (fishes with a bony skeleton). Instead of fleshy lobes, ray-fins have straight bony rays jutting from the body to support their fins. Modern forms have an all-bony skeleton; short, widely gaping jaws; thin scales; mobile fins for precise body control a symmetrical tail; and a lung evolved into a swim bladder to control buoyancy. These features evolved progressively through three great groups: first chondrosteans, then holosteans, then teleosts.

Early ray-fins were small species living about 370 million years ago. Later came also larger species, many living in the sea where they ousted placoderms. Fossil ray-fins occur worldwide: chondrosteans mostly in Devonian–Triassic rocks; holosteans mostly in Triassic–Cretaceous rocks; teleosts mostly in Jurassic and later rocks.

Fish in a fish
This remarkable fossil find from Texas shows a 4.3m (14ft) long *Xiphactinus* that swallowed a smaller relative 100 million years ago. Both were teleosts (advanced bony fishes) living in a sea that covered the southwest and south of the United States.

These three fishes represent evolutionary trends in the subclass Actinopterygii (ray-fins).

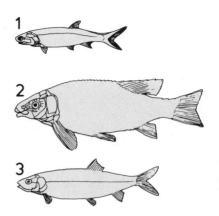

1 **Palaeoniscum** had a long upper tail lobe, thick scales, bony head armour, long jaws hinged far back, and a kind of lung. Length: 30cm (1ft). Time: mainly Permian. Place: worldwide. Infraclass: Chondrostei.

2 **Lepidotes** differed from *Palaeoniscum* in its short upper tail lobe, thinner scales, swim bladder, deeper body, more manoeuvrable paired fins, and shorter jaws. Length: up to 1.2m (4ft). Time: mainly Jurassic. Place: worldwide. Infraclass: Holostei.

3 **Leptolepis** had a herring-like shape. It differed from *Palaeoniscum* amd *Lepidotes* in its symmetrical tail, thinner scales, shortened jaws with wide gape, and fewer skull bones. Length: 23cm (9in). Time: mainly Jurassic. Place: seas worldwide. Infraclass: Teleostei.

Evolving tails and heads
Here we show evolutionary trends in bony fishes, from chondrosteans (**a,d**) through holosteans (**b,e**) to teleosts (**c,f**).
a Heterocercal tail: the backbone's upturned end produces a long lower tail lobe that tends to drive the head down.
b Abbreviated heterocercal tail: upper lobe shortened; lift comes from swim bladder.
c Homocercal tail: tail lobes seem equal, but most rays sprout from the backbone's still upturned end.
d Jaws work as a snap trap.
e Jaws are shortened but gape wide to suck in food.
f Jaws are shortened further but protrude when opened to create a suction tube.

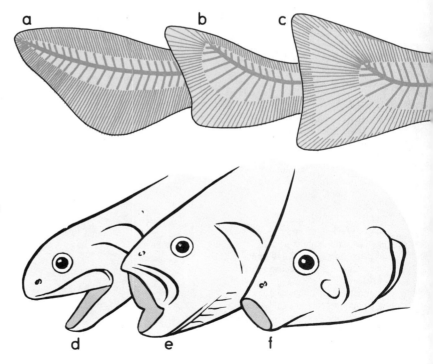

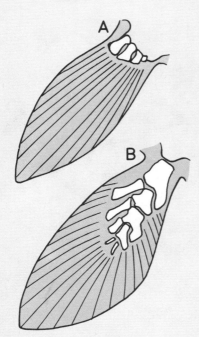

Bony fishes 2

The Sarcopterygian subclass of bony fishes had paired fins borne on scaly lobes containing bones and muscles. Such fleshy fins gave rise to the limbs of backboned animals that live on land. Fleshy-finned fishes appeared perhaps 390 million years ago. Three main groups evolved: rhipidistians, coelacanths, and lungfishes. The last two still survive.

Rhipidistians were long-bodied flesh-eaters that lurked in shallow waters – fresh and salt. They could use fins as legs and breathed air with lungs if hot weather made their water foul.

Coelacanths were mostly deep-bodied and lived in oceans. Their lungs became swim bladders that regulated buoyancy. All were thought extinct for over 60 million years until a fisherman caught one off south-east Africa in 1938.

Lungfishes had weaker limbs and a flimsier skeleton than other fleshy-finned fishes. Some could (and can) breathe atmospheric air if their ponds or rivers dry up.

Two types of fin (above)
A Ray-fin's fin: rays spring from bones at the base.
B Fleshy-finned fish's fin: rays spring from bones along the centre of the fin itself.

Fleshy-finned features (right)
Here we show similarities and differences between *Osteolepis* (**A**), an early fleshy-finned fish, and *Cheirolepis* (**B**), an early ray-finned fish. Both date from Mid Devonian times.
Similarities:
a tapered at both ends;
b covered with heavy scales;
c primitive, uptilted tail;
d paired fins similarly spaced;
e bony plates covering skull.
Differences:
f position and size of eyes;
g proportions of skull bones;
h number of dorsal fins;
i fin design.

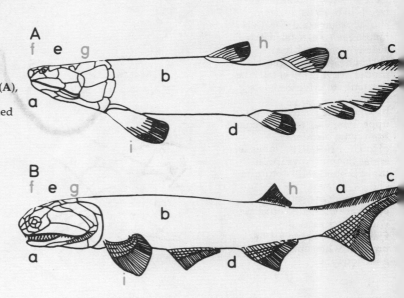

The animals below represent three main groups of fleshy-finned fishes. The first two were in the order Crossopterygii, the third in the order Dipnoi.

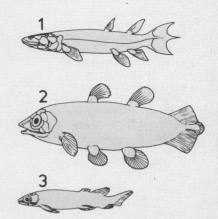

1 **Eusthenopteron** was a long-bodied, carnivorous freshwater fish with paired fins and a "three-pronged" tail fin. Skull, backbone, and limb bones resembled those of early amphibians. Nostrils opened into the mouth. Length: 30–60cm (1–2ft). Time: Late Devonian. Place: Europe and North America. Suborder: Rhipidistia.

2 **Macropoma** was a deep-bodied coelacanth with a short, deep skull, three-pronged tail fin, and fan-shaped dorsal and anal fins. Nostrils did not open into the mouth. Length: 56cm (22in). Time: Late Cretaceous. Place: oceans; fossils come from Europe. Suborder: Coelacanthini.

3 **Dipterus** had a long body tapered at both ends, paired, leaf-shaped fins, an uptilted tail, big, thick scales, and a braincase largely made of gristle. Length: 36cm (14in). Time: Middle Devonian. Place: North America and Europe. Order: Dipnoi (lungfishes).

Finny adventurers
Agile young of *Eusthenopteron* might have flipped ashore to dodge cannibal adults. Some found a food supply and stayed awhile. Breathing posed no problem, for these fishes were equipped with lungs.

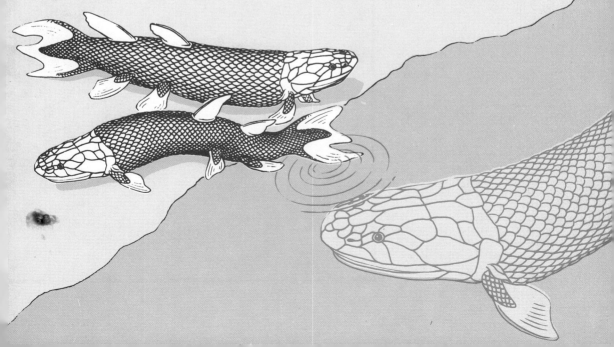

Chapter 5   FOSSIL AMPHIBIANS

The first vertebrates with legs were members of the class Amphibia. In this chapter we summarize their origins and key features, and show fossil examples of their four subclasses. These comprise the extinct labyrinthodonts (Labyrinthodontia) and lepospondyls (Lepospondyli), and the surviving frogs and toads (Anura), newts and salamanders (Urodela), and caecilians (Apoda).

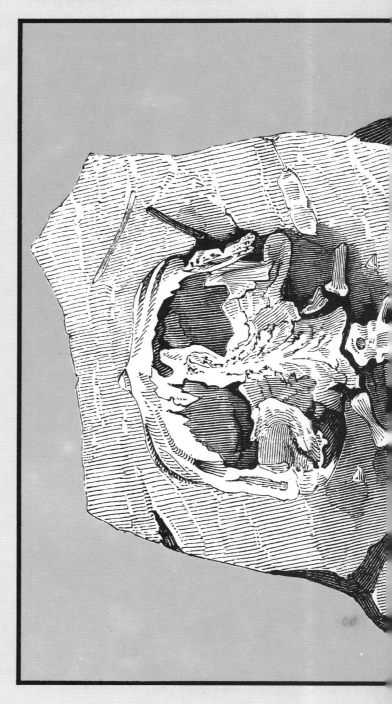

Andrias scheuchzerii was a Miocene salamander about 60cm (2ft) long. Impressed by its man-like form, Swiss scientist Johannes Scheuchzer in 1726 named the fossil *Homo diluvii testis* ("a man who had witnessed the flood"). Scheuchzer believed all fossils were remains of animals drowned by the biblical flood. Scientists now know that *Andrias scheuchzerii* was closely related to the living giant salamanders of China and Japan. (The Mansell Collection.)

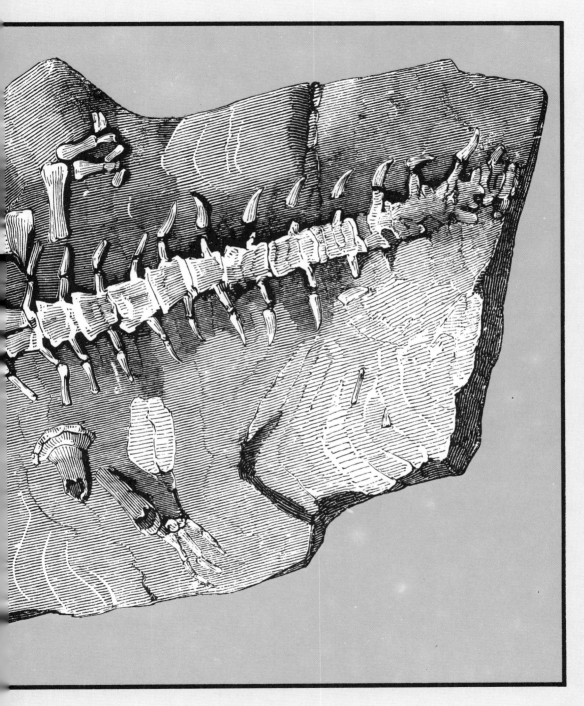

About amphibians

Rhipidistian fishes (p. 86) gave rise to amphibians about 350 million years ago. Amphibians were the first backboned animals with limbs designed for use on land. Like their descendants, frogs and newts, early amphibians had to lay their shell-less eggs in water to prevent them drying up. The eggs hatched into tadpoles which breathed through gills that later usually shrank. Some adults grew as big as any crocodile. Many had heavy skeletons, with powerful sprawling legs. They breathed through lungs and had a covering of fish-like scales or tough skin. (In contrast, modern amphibians are mostly small, with light skeletons and soft, moist skin; adults get more oxygen through skin than lungs.) Early amphibians lived mainly in or near fresh water, hunting fishes, insects, or early reptiles. They dominated swamps that covered much of coastal North America and

Amphibian family tree
In this family tree of the class Amphibia (amphibians) numbers show subclasses and superorders and letters show orders.
1 Labyrinthodontia (labyrinthodonts)
a Ichthyostegalia (ichthyostegids)
b Batrachosauria (batrachosaurs)
c Temnospondyli (temnospondyls)
2 Lepospondyli (lepospondyls)
d Nectridea (nectrideans)
e Aistopoda (aistopods)
f Microsauria (microsaurs)
3 Apoda (caecilians)
4 Anura (frogs and toads)
5 Urodela (newts and salamanders)

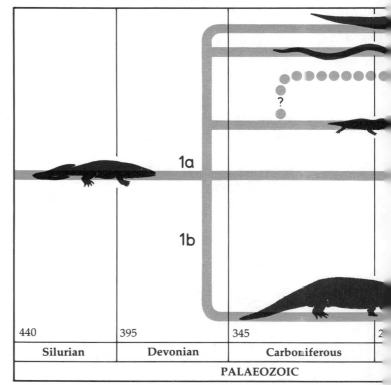

440	395	345	2
Silurian	Devonian	Carboniferous	
PALAEOZOIC			

© DIAGRAM

90

Europe in Late Carboniferous (Pennsylvanian) and Early Permian times. Fossils of extinct amphibian groups come mostly from rocks that formed then.

Both main groups of early amphibians died out by 160 million years ago. Meanwhile an unknown species of amphibian had given rise to reptiles and so, indirectly, to all other backboned land animals.

Limbs from fins
This comparison between bones of a fish's fin and an early amphibian's limb reveals that limb bones correspond to and evolved from fishes' fin bones.
1 Bones supporting a pelvic fin of a Devonian rhipidistian fish, *Eusthenopteron*
2 Corresponding bones in a hind limb of the Permian amphibian *Trematops*
a Pelvis (hip region)
b Femur (thigh bone in land vertebrates)
c Tibia and fibula (leg bones in land vertebrates)
d Pes (foot): the small bones in the rhipidistian fish's fins evolved into the toes and fingers of amphibians and their descendants
(Diagram after Colbert.)

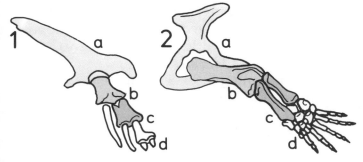

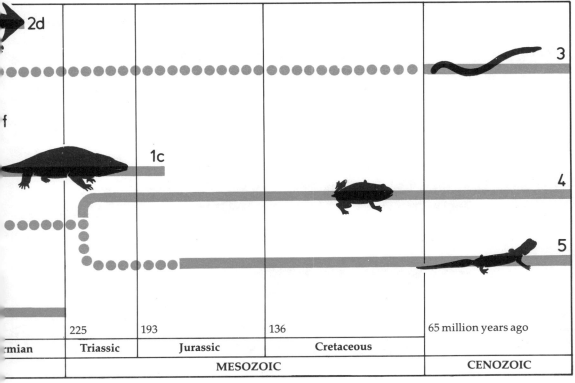

Labyrinthodonts 1

Labyrinthodonts formed the largest subclass of prehistoric amphibians. Their name comes from the labyrinth-like structure of their teeth.

Labyrinthodonts had solid skulls and complex spinal bones. Some evolved strong backbones and strong, sprawling limbs. These became the first land-living vertebrates. Others had weaker skeletons and eel-like bodies, and lived in water.

Labyrinthodonts ranged from a few centimetres long to 9m (30ft). They lived about 350–180 million years ago and spread worldwide. There were three orders: ichthyostegids, batrachosaurs (notably anthracosaurs), and (pp. 94–95) temnospondyls.

Ichthyostegids were sprawling amphibians with some fish-like features. Maybe they were ancestral to all later labyrinthodonts. Fossils come from Late Devonian rocks. (See example 1.)

Anthracosaurs were Late Palaeozoic amphibians. Some evolved like reptiles and might have given rise to these, but the anthracosaurs' middle ear design makes this unlikely. (See examples 2–6.)

1 **Ichthyostega** had well-developed limbs but traces of a fish's tail and scales. Length: 1m (3ft 3in). Time: Late Devonian. Place: Greenland.

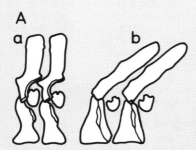

An early amphibian
Vertebrae (above) and skeleton and restoration (below) show features of *Ichthyostega*.
Fish-like features are:
A bones of vertebrae (**a**) matching those found in the fish *Eusthenopteron* (**b**);
B skull roof still solid;
C fish-like tail;
D fish-like scales.
Amphibian innovations are:
E strong shoulder girdle;
F strengthened spine;
G strong ribs
H strengthened hip girdle;
I fully formed limbs.

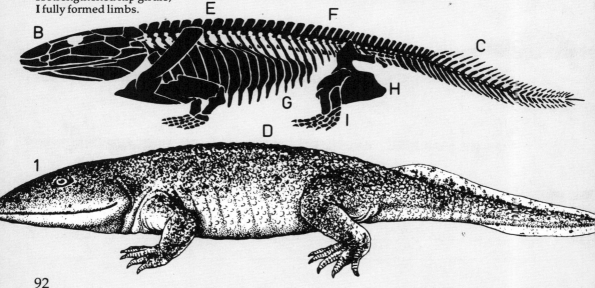

2 **Protogyrinus,** a very early anthracosaur, had a rather high skull and sturdy limbs. Length: 1–1.5m (3ft 3in–4ft 11in). Time: Late Carboniferous. Place: West Virginia, USA. Infraorder: Embolomeri.
3 **Gephyrostegus** had a small head and sturdy, sprawling limbs. Length: 45cm (18in). Time: Late Carboniferous. Place: Czechoslovakia. Infraorder: Gephyrostegoidea (terrestrial anthracosaurs).
4 **Eogyrinus** had a long, eel-like body and tail, weak limbs, and a crocodile-like skull. It lived in water. Length: 4.6m (15ft). Time: Late Carboniferous. Place: Europe. Infraorder: Embolomeri ("typical" aquatic anthracosaurs).
5 **Seymouria** had longer, stronger limbs than the first amphibians and lived on land. Length: 60cm (2ft). Time: Early Permian. Place: Texas, USA. Infraorder: Seymouriamorpha (reptile-like anthracosaurs).
6 **Diadectes,** the earliest known plant-eating vertebrate, had heavy bones and shortened jaws with blunt teeth. Length: 3m (10ft). Time: Early Permian. Place: Texas, USA. Infraorder: Seymouriamorpha.

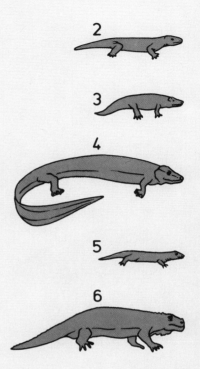

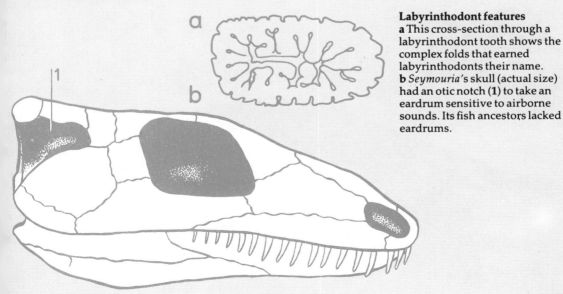

Labyrinthodont features
a This cross-section through a labyrinthodont tooth shows the complex folds that earned labyrinthodonts their name.
b *Seymouria's* skull (actual size) had an otic notch (**1**) to take an eardrum sensitive to airborne sounds. Its fish ancestors lacked eardrums.

©DIAGRAM

Labyrinthodonts 2

Temnospondyl labyrinthodonts had distinctive
vertebrae and other features that distinguish
them from anthracosaurs. Like those, some
temnospondyls lived on land, others in water. They
persisted from Carboniferous to Jurassic times and
became the most abundant amphibians.

The temnospondyl order held three suborders:
rhachitomes, stereospondyls, and plagiosaurs.
Rhachitomes, the basic stock, formed a large, varied
group: dominant amphibians of Permian times.

1 **Eryops** was heavy with a strong skeleton, short,
strong limbs, and big, broad skull. It might have
lived in water and on land, like crocodiles. Length:
1.5m (5ft). Time: Early Permian. Place: Texas, USA.

2 **Trimerorhachis** had a body protected by
overlapping "fish scales". It probably swam in pools
and streams. Length: 60cm (2ft). Time: Early
Permian. Place: Texas, USA.

3 **Aphaneramma** had a head one third of its total
length. It swam in seas and caught fishes with its
long, slim, sharp-toothed jaws. Length: 60cm (2ft).
Time: Early Triassic. Place: worldwide. Fossils occur
as far apart as Australia and Spitsbergen.

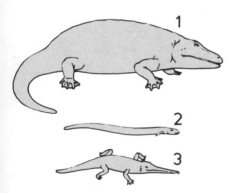

A landlubber (below)
The Permian temnospondyl
Cacops was one of the
amphibians best designed for
life on land. It had these features.
a Length 40cm (16in)
b Sturdy limbs
c Large eardrum
d Armoured skin on the back
e Large head with long jaws

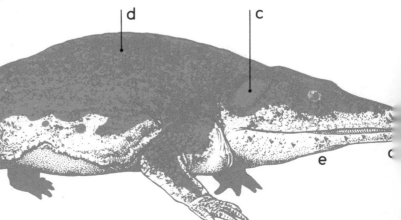

94

Stereospondyls were swimmers with degenerate skeletons. They did not need strong bony scaffolding like beasts that live on land and must resist the tug of gravity. Some developed broad, flat bodies and huge heads. Stereospondyls included the largest amphibians ever. The group dominated inland waters in Triassic times, then all died out.

Plagiosaurs were even more grotesque aquatic beasts, of Permian to Late Triassic times.

1 **Cyclotosaurus,** a stereospondyl, was as large as a crocodile but it had very small, weak legs and needed the support of water. Length: 4.3m (14ft). Time: Late Triassic. Place: Europe.

2 **Gerrothorax** was a plagiosaur with a short, wide head with gills; a flat, broad, armoured body; short tail; and tiny limbs. Length: 1m (3ft 3in). Time: Late Triassic. Place: southern Germany.

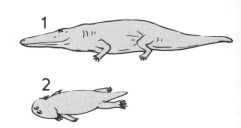

Built for water life (below)
Paracyclotosaurus, an aquatic Triassic stereospondyl, was modified for life in water.
a Length 2.25m (7ft 5in)
b Short, rather weak limbs
c Somewhat flattened body
d Flattened skull
e Mouth opened by raising the skull instead of dropping the lower jaw

©DIAGRAM

Lepospondyls

Lepospondyls formed a mixed subclass of small amphibians that thrived in swamps about 320–230 million years ago, dying out before Triassic times. They had simpler teeth and fewer skull bones than labyrinthodonts, and a different type of vertebra. This had a bony cylinder firmly joined to an arch and pierced by a hole to take a notochord. Below are examples of the three lepospondyl orders: aistopods, nectrideans, and microsaurs.

1 **Ophiderpeton** was a typical aistopod: aquatic, limbless, and snake-like, with forked ribs and 200 vertebrae. Length: 70cm (27.5in). Time: Late Carboniferous. Place: Europe and North America.

2 **Diplocaulus** had a flat body, weak limbs, and a grotesque head like a cocked hat. Length: 1m (3ft 3in). Time: Early Permian. Place: Texas, USA. Order: Nectridea (newt-like or snake-like).

3 **Pantylus** had a heavy body, small limbs, and a big, deep head. Length: 26cm (10in). Time: Early Permian. Place: Texas, USA. Order: Microsauria (mostly sturdy, land-based insect-eaters)

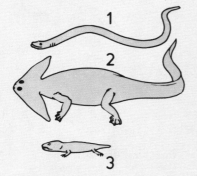

A puzzling head
Diplocaulus's strange head poses questions of design.
A Growth stages show that "horns" grew relatively longer as the head enlarged. This occurred because some skull bones outpaced others.
B Two illustrations suggest possible uses for a head with backswept "horns".
1 "Horns" might have acted as a hydrofoil, helping to give lift to raise the animal through the water.
2 "Horns" might have made it difficult for predators to swallow *Diplocaulus*.

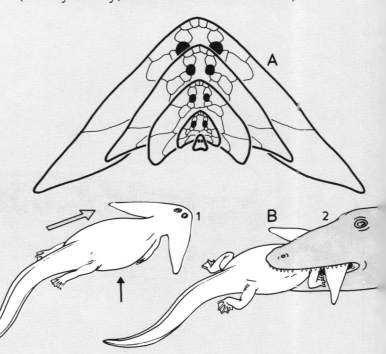

Modern amphibians

Frogs and toads, tailed amphibians, and caecilians are often lumped together in one subclass: Lissamphibia. Few known ancestors of modern amphibians date back to the last labyrinthodonts but similarities in skull design show that frogs, toads and urodeles came from a group of temnospondyl labyrinthodonts (the dissorophids). Below are early examples of each group.

1 **Triadobatrachus,** the first known fossil frog, lived in Early Triassic Madagascar. Length: 10cm (4in). Superorder: Anura.

2 **Karaurus,** the oldest known complete salamander skeleton, had a broad skull with sculptured bones. Length: 19cm (7.5in). Time: Late Jurassic. Place: Kazakhstan, USSR. Superorder: Urodela (newts and salamanders).

3 **Apodops,** an early apodan, is known from one vertebra. Time: Palaeocene. Place: Brazil. Superorder: Apoda (caecilians – worm-like burrowers).

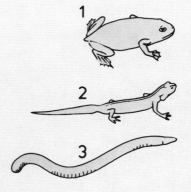

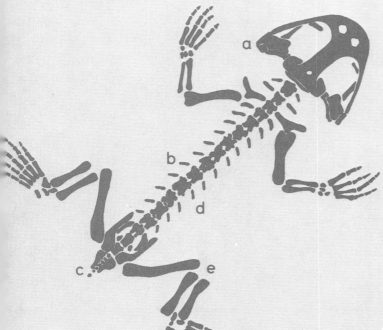

A frog ancestor
Here we show evolutionary changes that produced *Triadobatrachus* from long-bodied amphibian ancestors.
a Frog-like skull
b Shortened back with reduced spinal bones (modern frogs have fewer still)
c Shortened tail (modern adult frogs have none)
d Shortened ribs (modern frogs have none)
e Leg design still primitive (modern frogs have very long hind limbs, for jumping)

Chapter 6

FOSSIL REPTILES

The class Reptilia includes the first vertebrates designed to live and also to breed on land. Reptiles colonized the continents, and for 200 million years were masters over them.

These pages explore key features and fossil examples of orders belonging to the four reptile subclasses: the extinct euryapsids (Euryapsida) and mammal-like synapsids (Synapsida), and the surviving anapsids (Anapsida) and diapsids (Diapsida).

Limbs that evolved as flippers adapted *Plesiosaurus* for life in water. This large aquatic reptile flourished in the Mesozoic Era: the "Age of Reptiles" or "Age of Dinosaurs". Our book follows the practice of classifying plesiosaurs in one of four reptile subclasses determined by holes in the skull. A classification proposed in 1980 would restructure subclasses and regroup their contents. (Illustration from *A History of British Fossil Reptiles* by Sir Richard Owen.)

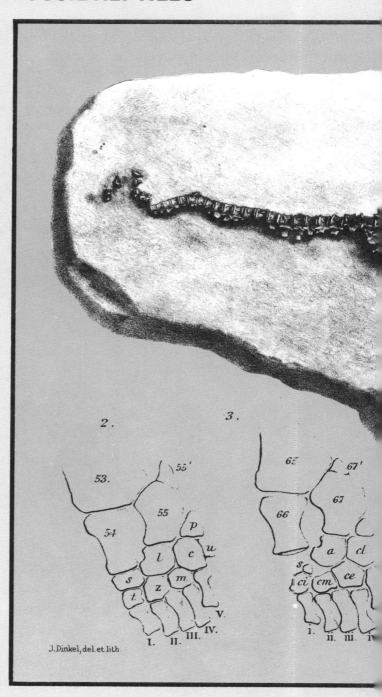

J.Dinkel, del. et lith

PLESIOSAURUS RUGOSUS

About reptiles

By 300 million years ago amphibians had given rise to reptiles, the first backboned animals able to live entirely on dry land. Reptiles are cold-blooded animals with dry, scaly, waterproof skin. Their eggs are fertilized inside the females. Tough skin or a hard shell stops the eggs drying up after laying. These devices freed reptiles from the waterside. They colonized high, dry lands between river valleys and invaded the spreading deserts of Permian and Triassic times.

By 270 million years ago small, early, swampland reptiles had given rise to four great stocks named from the number and type of holes behind the eyes, on each side of the skull – holes that left space for strong jaw muscles to contract. Anapsids, without such holes, possibly include tortoises and turtles. Synapsids, with one low opening in the cheek, made

Reptile family tree
This shows major reptile groups as set out in this chapter. (New studies will lead to some revision.) Numbers represent subclasses: small letters represent orders.

1 Anapsida (anapsids)
a Cotylosauria (cotylosaurs)
b Mesosauria (mesosaurs)
c Chelonia (turtles)
2 Synapsida (synapsids)
d Pelycosauria (pelycosaurs)
e Therapsida (therapsids)
3 Euryapsida (euryapsids)
f Ichthyosauria (ichthyosaurs)
g Araeoscelidea (araeoscelids)
h Placodontia (placodonts)
i Sauropterygia (nothosaurs and plesiosaurs)
4 Diapsida (diapsids)
A Infraclass Lepidosauria (lepidosaurs)
j Eosuchia (eosuchians)
k Rhynchocephalia (rhynchocephalians)
l Squamata (lizards and snakes)
B Infraclass Archosauria (archosaurs)
m Thecodontia (thecodonts)
n Crocodilia (crocodilians)
o Saurischia (saurischian dinosaurs)
p Ornithischia (ornithischian dinosaurs)
q Pterosauria (pterosaurs)

345	280	225	19
Carboniferous	Permian	Triassic	
PALAEOZOIC			

up the mammal-like reptiles. These dominated life on land for 70 million years and then became extinct. Meanwhile some gave rise to mammals. Euryapsids, with a high opening in the cheek, were mostly sea reptiles such as plesiosaurs; none survives. Diapsids, with two openings behind each eye, had two main subgroups: lepidosaurs and archosaurs. Lepidosaurs included snakes, lizards, and their ancestors. Archosaurs comprised the thecodonts, crocodiles, dinosaurs, and pterosaurs.

From Late Palaeozoic times all through the Mesozoic Era, big reptile predators and herbivores were the "lions" and "zebras" of their day. Others evolved wings or flippers, and took to air or water. Worldwide fossil finds bear witness to this Age of Reptiles. Such beasts no longer rule the Earth, but their heirs the birds and mammals do.

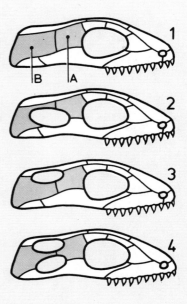

Four types of reptiles
Here reptiles are put in four subclasses according to their skull design (though some show exceptions to this rule and a 1980 reclassification proposes a different breakdown).
1 Anapsids show no hole between postorbital (**A**) and squamosal (**B**) bones.
2 Synapsids show one hole between and below these bones.
3 Euryapsids show one hole between and above these bones.
4 Diapsids show two holes between these bones.

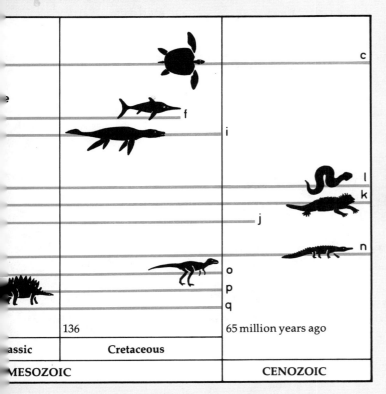

			c
		f	
		i	
		l	
		k	
		j	
		n	
		o	
		p	
		q	
	136	65 million years ago	
...assic	Cretaceous		
MESOZOIC		CENOZOIC	

Reptile pioneers

Cotylosaurs ("stem reptiles") formed the first known reptile order. Cotylosaurs ranged from small, lizard-like beasts to big, bulky creatures 3m (10ft) long. They had sturdy, sprawling limbs and a solidly roofed skull, with eardrums just above the jaw hinge. Most inhabited swamps, 290–200 million years ago. Fossils crop up worldwide, mostly in Permian and Triassic rocks.

Mesosaurs were very early swimming reptiles that caught fishes in freshwater ponds and lakes. Fossils some 265 million years old come from Lower Permian rocks of Brazil and southern Africa. Cotylosaurs probably gave rise to mesosaurs and all the other reptile orders.

We show here a mesosaur and examples of both cotylosaur suborders: captorhinomorphs (small, carnivorous reptiles) and procolophonoids (the captorhinomorphs' mostly small descendants).
1 **Hylonomus,** one of the first reptiles, was a small, low captorhinomorph, with sprawling limbs, long tail, short neck, short, pointed snout, and sharp teeth. Length: 1m (3ft 3in). Time: Late Carboniferous (Lower Pennsylvanian). Place: Nova Scotia, Canada.

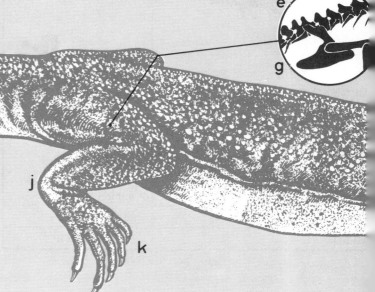

2 **Hypsognathus** was a lizard-like procolophonian, with broad cheek teeth, and spikes jutting back from its head. Length: 33cm (13in). Time: Late Triassic. Place: New Jersey, USA.

3 **Scutosaurus** belonged to the pareiasaurs: big, heavy, procolophonoid herbivores, that stood more upright than many reptiles. The small, saw-edged teeth in its broad head probably sliced up vegetation. Length: 2.4m (8ft). Time: Late Permian. Place: USSR.

4 **Millerosaurus** was a lizard-like reptile, with broad cheeks, short jaws, and a skull hole behind each eye. It might have given rise to the diapsids. Length: 1m (3ft 3in). Time: Late Permian. Place: South Africa.

5 **Mesosaurus,** a mesosaur, had a long, slim body, slender, sharp-toothed jaws, paddle-shaped limbs, and long, deep, swimmer's tail. Length: 71cm (28in). Time: Early Permian. Place: Brazil and South Africa.

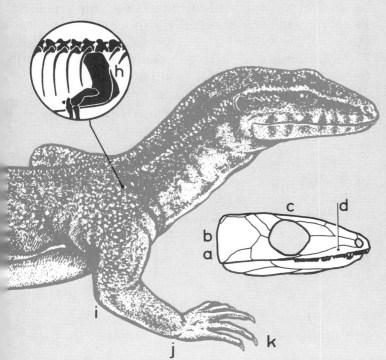

Reptilian features
Letters indicate features that help palaeontologists to identify fossil bones as those of a reptile, not an amphibian, though certain amphibians share some of these features.
a Relatively deep skull
b Usually no otic (ear) notch
c Small size or distinctive position of some skull bones
d Any teeth on palate small not tusk-like
e Two or more vertebrae join spine to hip girdle
f Pleurocentrum is the main part of each vertebra
g Enlarged ilium (a hip bone)
h Shoulder blade well developed but some bones in the shoulder girdle reduced
i Limb bones slimmer than in labyrinthodont amphibians
j Fewer wrist and ankle bones than in amphibians
k Distinctive numbers of toe and "finger" bones

©DIAGRAM

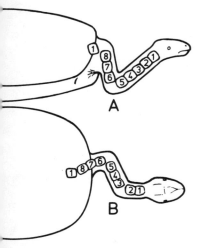

Turtles and tortoises

Chelonians (turtles and tortoises) form one of the oldest living reptile orders. Probably evolving from a cotylosaur, they appeared well developed by 205 million years ago.

Most had a broad, short body protected by a bony shell sheathed with horny scutes. Many could (and can) pull head, tail, and limbs inside the shell for protection. They evolved a toothless beak for slicing meat or vegetation. From the first chelonians came today's land tortoises, and turtles designed to swim and hunt in rivers, pools, or seas. Of four (perhaps five) suborders, two survive.

The earliest fossils come from Triassic Germany and Thailand. Later fossils occur worldwide.

We show three chelonians representing evolutionary trends: proganochelyids gave rise to amphichelyids (not shown) which led to cryptodires and pleurodires (both flourishing today).

1 **Proganochelys** had a well-developed shell but "old-fashioned" skull with teeth as well as beak. Probably it could not pull limbs, tail, or head inside its shell. Shell length: 61cm (2ft). Time: Late Triassic. Place: Germany. Suborder: Proganochelydia (ancestral chelonians).

2 **Podocnemis** was and is a freshwater turtle (terrapin) pulling the neck in sideways. Length: to 76cm (30in). Time: Late Cretaceous onward. Place: once widespread, now South America and Madagascar. Suborder: Pleurodira (side-neck turtles).

3 **Archelon** was a huge, early marine turtle. It had a broad, light, flattened shell and long paddle-like limbs. Length: 3.7m (12ft). Time: Late Cretaceous. Place: North America. Suborder: Cryptodira (vertical-neck turtles, including all living tortoises and most turtles).

Neck benders (above)
These diagrams reveal how two types of chelonian pull the head inside the shell.
A Cryptodires bend the neck down and back. Bending occurs mostly between cervical (neck) vertebrae 5 and 6, and between cervical vertebra 8 and dorsal (back) vertebra 1.
B Pleurodires bend the neck sideways. Bending occurs mostly between cervical vertebrae 2 and 3 and 5 and 6, and between cervical vertebra 8 and dorsal vertebra 1.

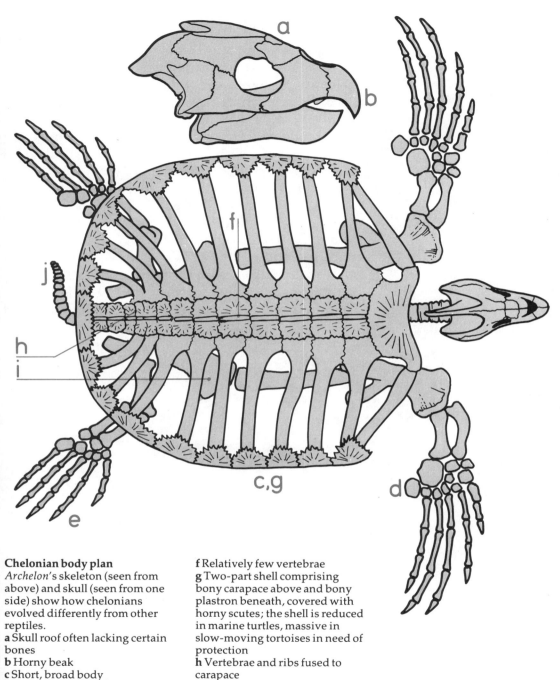

Chelonian body plan

Archelon's skeleton (seen from above) and skull (seen from one side) show how chelonians evolved differently from other reptiles.

a Skull roof often lacking certain bones
b Horny beak
c Short, broad body
d Heavy limbs projecting sideways
e Relatively few toe bones

f Relatively few vertebrae
g Two-part shell comprising bony carapace above and bony plastron beneath, covered with horny scutes; the shell is reduced in marine turtles, massive in slow-moving tortoises in need of protection
h Vertebrae and ribs fused to carapace
i Limb girdles and upper limb bones fitting inside ribs
j Short tail

Early euryapsids

Euryapsids were a subclass of mostly marine reptiles with a hole high in each temple of the skull. Three major early groups were the araeoscelids, placodonts, and nothosaurs. Araeoscelids were lizard-like and largely lived on land. Placodonts had short, stout, armoured bodies, paddle-like limbs, and blunt teeth. Nothosaurs were slimmer with longer necks and bodies, and sharp teeth. Both groups hunted in the sea: placodonts for molluscs, nothosaurs for fishes. Araeoscelids seemingly evolved from primitive cotylosaur reptiles in early Permian times, 270 million years ago. Placodonts and nothosaurs flourished in Triassic times until wiped out by competition from bony fishes and new aquatic reptiles. Many fossils in these groups occur in Triassic rocks in parts of Europe, North Africa, and Asia that rimmed the ancient Tethys Sea.

Placodont body plan
The 2m (6ft 6in) *Placodus* from Mid Triassic Europe showed these typical placodont features.
a Powerful jaws
b Peg-like front teeth
c Flat, broad, crushing tooth plates at the back of the mouth
d Short, heavy, rounded body
e Short neck
f Extra set of (belly) ribs
g Small bones forming protective armour
h Limbs designed as paddles
i Skin joining toes and fingers
j Flattened tail (short in later placodonts)

1 **Araeoscelis** was small, lightly built, and rather lizard-like, but with long, slim shins and "forearms". Length: 66cm (26in). Time: Early Permian. Place: Texas, North America. Order: Araeoscelidea.

2 **Tanystrophaeus** had a grotesquely long neck, used maybe as a fishing rod. Adults lived in the sea, young on shore. Length: up to 6m (20ft). Time: Mid Triassic. Place: Central Europe and Israel. Order: Araeoscelidea, or maybe Eosuchia (p. 112).

3 **Henodus** shows how later placodonts evolved like marine turtles: with flippers, a broad, flat body protected by a bony shell, and a horny, toothless beak. Length: 1m (3ft 3in). Time: Late Triassic. Place: Germany. Order: Placodontia.

4 **Nothosaurus** had a long, slim neck and body, long forelimbs, and long, slim jaws bristling with sharp teeth shaped for catching fishes. Length: 3m (10ft). Time: Mid Triassic. Place: Central Europe, North Africa, South-West Asia, East Asia. Order: Sauropterygia. Suborder: Nothosauria.

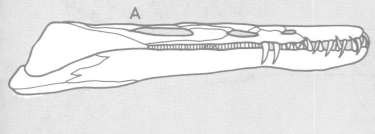

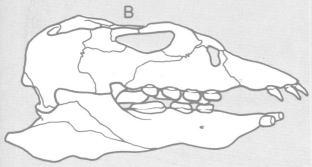

Skulls compared
These diagrams contrast the jaws of two euryapsids that ate different foods.
A *Nothosaurus* jaws were long and slim, with long, sharp teeth that interlocked to seize and grip slippery fishes.
B *Placodus* jaws were deep and strong, and closed by muscles with great crushing force. Jutting front teeth were pincers to pluck molluscs from the sea bed. Flat back teeth and bones in the skull roof and lower jaw crushed mollusc shells.

Plesiosaurs

Plesiosaurs ("near lizards") were bigger than the Triassic nothosaurs and better built for water life. They had a bulky, barrel-shaped body; broad ribs; "belly" ribs; four long, flat flippers; and a short tail. With ichthyosaurs they ruled Jurassic and Cretaceous seas and oceans, and then died off.

Their suborder held two superfamilies. Long-necked plesiosaurs were expert fishers at or near the surface. Short-necked plesiosaurs (pliosaurs) dived and preyed on ammonites. Both types swam with flippers, as marine turtles do. Flippers also helped haul them ashore to lay eggs.

Fossil plesiosaurs occur in Jurassic and Cretaceous clays and limestones – especially Liassic (early Jurassic) rocks in England and Germany, Late Cretaceous rocks in western North America, and Early Cretaceous rocks in Australia.

Plesiosaur body plan
Long-necked plesiosaurs had the following features.
a Bulky, rounded body
b Very long front flippers
c Hind flippers shorter than front flippers
d Very long neck, for darting down on fishes from above or snaking swiftly sideways through the water
e Small head
f Sharp, needle-like teeth for gripping fishes
g Short tail

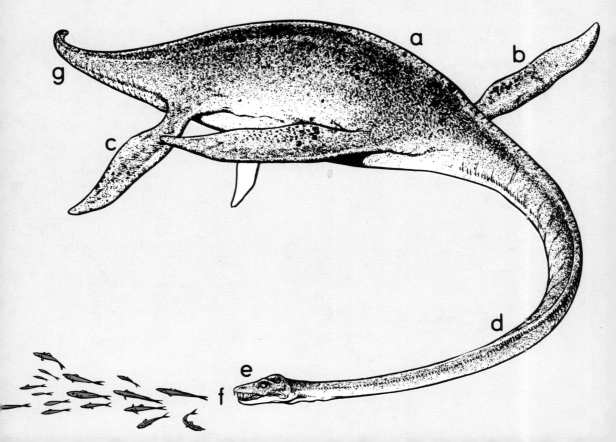

These four plesiosaurs represent evolutionary trends in both superfamilies.

1 **Peloneustes** was a small short-necked plesiosaur. Length: 3m (10ft). Time: Late Jurassic. Place: Western Europe. Superfamily: Pliosauroidea.

2 **Kronosaurus** was a huge short-necked plesiosaur with massive teeth. Length: up to 17m (56ft), one quarter of this being head. Time: Early Cretaceous. Place: Australia. Superfamily: Pliosauroidea.

3 **Thaumatosaurus,** an early long-necked plesiosaur, had a neck less than one quarter its total length. Length: 3.4m (11ft). Time: Early-Mid Jurassic. Place: Europe. Superfamily: Plesiosauroidea.

4 **Elasmosaurus** was more than half neck, with more than 70 neck vertebrae. Length: 12m (39ft) or longer. Time: Late Cretaceous. Place: North America. Superfamily: Plesiosauroidea.

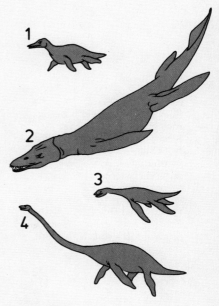

Plesiosaur skeleton
Seen from below, the skeleton of *Thaumatosaurus* shows the overall design of an early long-necked plesiosaur.
a Moderate length: 3.4m (11ft)
b Many neck bones
c Plate-like shoulder bones anchoring muscles that pull front flippers down and back
d Plate-like hip bones supporting muscles that operate hind flippers
e Short upper limb bones
f Long feet and "hands"
g Belly ribs

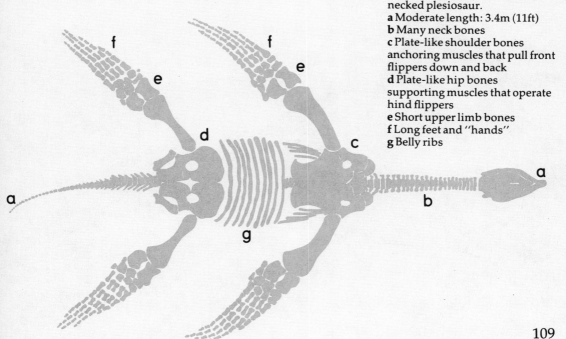

Ichthyosaurs

Ichthyosaurs ("fish lizards") were aquatic reptiles that flourished in shallow seas about 220 million to 90 million years ago. They had fins, flippers, long narrow jaws, and a superbly streamlined body. The largest individuals measured about 7.6m (25ft) but some species were no longer than a man. Ichthyosaurs ate cephalopods and fish, and gave birth to their young in water.

These reptiles spread all around the world, but their fossils are most plentiful in Lower Jurassic rocks about 180 million years old. Fossil ichthyosaur bones or faeces occur in, for instance, Britain's Lias shales and limestones, and in rocks in Germany, North and South America, Australia, and Indonesia.

Skin and bones
Fine-grained Lower Jurassic rock preserved this fine *Stenopterygius* ichthyosaur fossil from Boll Holzmaden in southern Germany. Even skin remains survive. The fossil's biconcave vertebrae, bony eye ring, and slim, toothy jaws are typical of ichthyosaurs. Experts familiar with such features can often identify an ichthyosaur from just one bone.

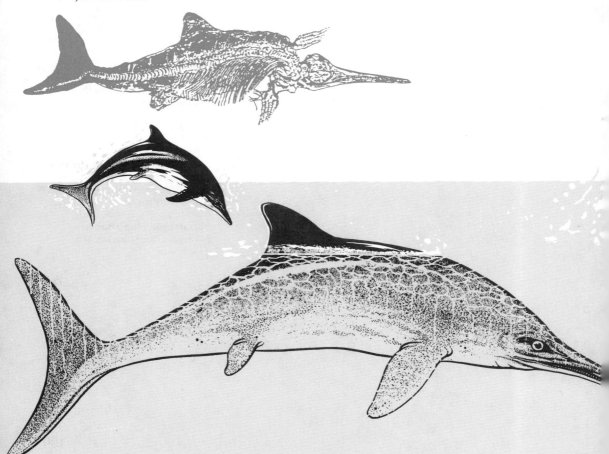

These three ichthyosaurs represent evolutionary trends in the order Ichthyosauria.

1 **Cymbospondylus,** a Triassic ichthyosaur, had relatively long arm and thigh bones, short skull, and small tail, recalling the body build of the ichthyosaurs' land-based ancestors.

2 **Ichthyosaurus,** from Lower Jurassic rocks, was much more streamlined than *Cymbospondylus*. It had a large dorsal fin and tail, forelimbs broadened into paddles, and a long, tapered skull.

3 **Ophthalmosaurus,** from Upper Jurassic rocks, reveals further evolutionary changes, including enlarged eyes and propulsive tail, tiny hind limbs, and teeth in a groove instead of separately socketed as in early ichthyosaurs.

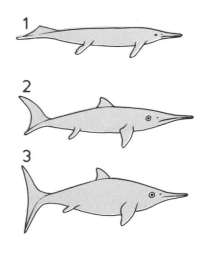

Dolphin's lookalike
Ichthyosaurs could have sheared swiftly through the waves. Their bodies were designed for speed, much like those of dolphins – living sea mammals evolved from ancestors that dwelt on land. But ichthyosaurs had vertical not horizontal tails, longer jaws, and simpler brains that made them less intelligent than dolphins.

©DIAGRAM

111

Eosuchians

Eosuchians ("dawn crocodiles") were mostly small, lizard-like creatures, although some of these gave rise to larger, crocodile-like reptiles. The order Eosuchia lasted from 290 to 50 million years ago, but their heyday was 280–200 million years ago. Fossils crop up mostly in Upper Permian rocks of South Africa, though finds occur elsewhere, especially in North America and Europe.

Eosuchians evolved from cotylosaurs ("stem reptiles") and were the first diapsids (reptiles of the subclass with two openings in the skull behind each eye). Their infraclass, Lepidosauria, includes the snakes and lizards.

1 **Youngina** had slim limbs, a long, slim tail and body, and a pointed skull of "old-fashioned" design, with teeth inside the mouth as well as set in sockets

Diapsid family tree
This family tree shows likely relationships of orders of the reptile subclass Diapsida.
1 Cotylosauria (stem reptile ancestors)
2 Eosuchia (eosuchians)
3 Rhynchocephalia (beak-heads or rhynchocephalians)
4 Squamata (lizards and their offshoot, snakes)
5 Thecodontia (thecodonts)
6 Crocodilia (crocodilians)
7 Saurischia (saurischian dinosaurs)
8 Ornithischia (ornithischian dinosaurs), derived from either thecodonts or saurischian dinosaurs
9 Pterosauria (pterosaurs)

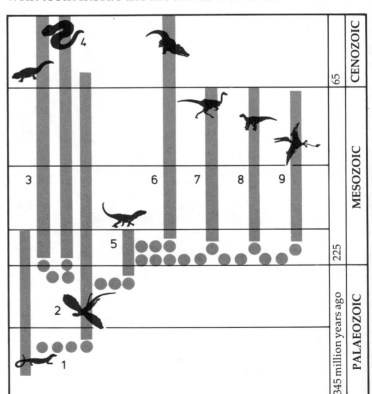

© DIAGRAM

112

on the jaw rim. Length: 45cm (18 in). Time: Late
Permian. Place: South Africa. Suborder:
Younginiformes (lizard-like).

2 **Weigeltisaurus** glided from tree to tree on skin
wings stretched between enormously long ribs.
Length: 50cm (20in). Time: Late Permian. Place:
England and Germany.

3 **Askeptosaurus** had a long, slim body, long
sharp-toothed skull, and small, paddle-like limbs.
Length: 1.5m (5ft). Time: Mid Triassic. Place:
Europe. Suborder: Thalattosauria (aquatic hunters).

4 **Champsosaurus** resembled a slim-snouted
crocodilian. Length: 1.5m (5ft). Time: Late
Cretaceous-Eocene. Place: North America and
Europe. Suborder: Choristodera (fish hunters).

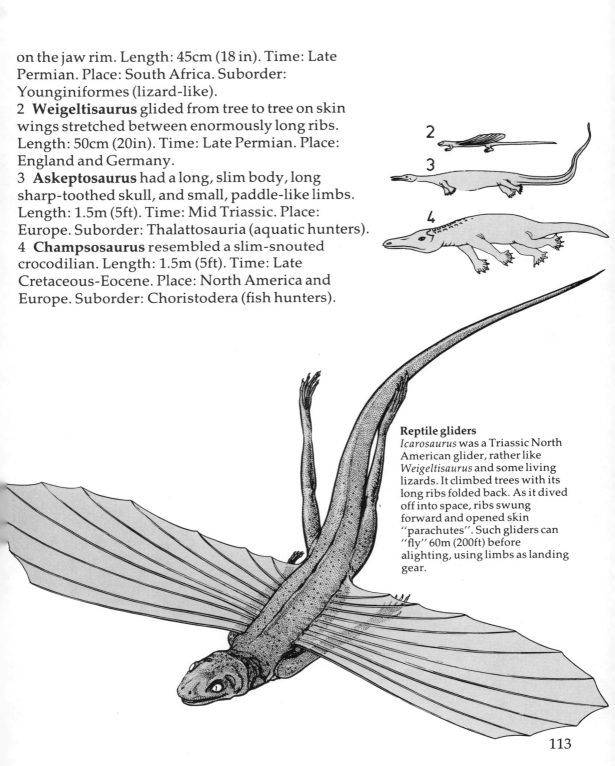

Reptile gliders
Icarosaurus was a Triassic North
American glider, rather like
Weigeltisaurus and some living
lizards. It climbed trees with its
long ribs folded back. As it dived
off into space, ribs swung
forward and opened skin
"parachutes". Such gliders can
"fly" 60m (200ft) before
alighting, using limbs as landing
gear.

113

Lizards and snakes

With their eosuchian ancestors, lizards, snakes, and rhynchocephalians ("beak-heads") form an infraclass, the lepidosaurs.

Lizards appeared about 230 million years ago and gave rise to some strange aquatic forms. Mosasaurs ("Meuse lizards") were a family of huge Late Cretaceous sea lizards that seized ammonites and fish in sharp-toothed jaws. Mosasaur fossils are especially plentiful in Niobrara chalk from Kansas.

By 70 million years ago, lizards had given rise to snakes. Between them, snakes and lizards make up the Squamata – the most numerous, varied, and widespread reptile order in the world today. Here are four prehistoric examples.

1 **Paliguana** was a small, early lizard, or maybe a lizard ancestor, from Early Triassic South Africa.

2 **Acteosaurus** had a long, skinny body, slim tail, and short limbs. It was in the dolichosaurid family of semi-aquatic Cretaceous European lizards. Length: about 40cm (16in).

3 **Tylosaurus** was a mosasaur up to 8m (26ft) long. Place: North America and New Zealand.

4 **Dinilysia** might have been an early relative of modern boas and pythons. Length: 1.8m (6ft). Time: Late Cretaceous. Place: Patagonia, South America.

Mosasaur body plan
a Huge size
b Long head
c Nostrils high on skull
d Long jaws, with a joint in the lower jaw
e Sharp teeth set in sockets
f Short neck
g Long body
h Flattened, paddle-shaped limbs to steer and balance
i Long tail flattened from side to side for swimming

Rhynchocephalians

Rhynchocephalians ("beak-heads") had an upper jaw ending in a down-curved beak. Some were lizard-like, some bulkier. They evolved and spread in the Triassic Period, then largely fizzled out. Only one species is alive today.

1 Homoeosaurus was a lizard-like Late Jurassic beast possibly related to the living tuatara found only in New Zealand. Length: 19cm (7.5in). Place: south-west Germany.

2 Scaphonyx belonged to the rhynchosaurid family of heavy-bodied beak-heads with a deep skull, toothless, tong-like jaws, and rows of crushing toothplates in the mouth. Probably it chopped and crushed up husked fruits. Length: 1.8m (6ft). Weight: about 90kg (200lb). Time: Mid Triassic. Place: Brazil. (Others lived in Africa and India.)

Clues to diet
Rhynchosaurs probably ate plants. Clues include features of skull and teeth, shown here from below (**A**) and side (**B**):
a rows of teeth on tooth plates;
b groove to take the lower jaw as jaws shut like a penknife:
c tong-like food-gathering beak, manipulated by a large tongue. Body restorations also give clues to diet. Strong hind limbs (**C**) could have scratched up roots and tubers. The "barrel body" (**D**) left room for a large gut, digesting bulky food.

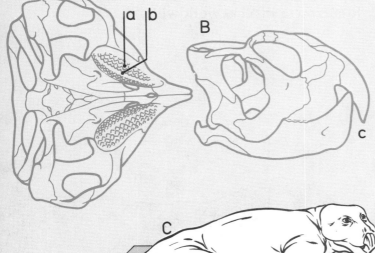

Thecodonts

Thecodonts ("socket toothed" reptiles) were a hugely important reptile order, for they gave rise to the crocodiles, dinosaurs, and pterosaurs. All four made up one infraclass: the archosaurs or "ruling reptiles". Archosaurs dominated life on land in Mesozoic times.

Thecodonts probably evolved from eosuchians about 226 million years ago. Most were large, four-legged flesh-eaters with distinctive skeletons. They came, perhaps, in four suborders. First were sprawling, heavy-bodied proterosuchians. From these sprang pseudosuchians, including small, light creatures with bodies held well off the ground. Pseudosuchians gave rise to four-legged, armoured herbivores called aëtosaurs, and to the flesh-eating, crocodile-like phytosaurs.

Thecodonts died out about 193 million years ago, replaced by dinosaurs and crocodiles. Thecodont fossils occur in rocks worldwide.

Two ways of standing (above)
A Many thecodonts sprawled, knees and elbows stuck out and feet flat on the ground.
B Pseudosuchians tucked the knees and elbows down and in to lift the body and gain speed. Some ran on their toes.

Thecodont body plan (right)
A typical thecodont had these features.
1 Skull lightened on each side by two holes behind the eye (**a**), one in front of the eye (**b**), and one in the lower jaw (**c**).
2 Teeth set in sockets
3 About two dozen vertebrae between head and hip region
4 Two sacral vertebrae (linked to the hips)
5 Hip socket shaped as a solid bony basin
6 Fairly straight thighbone, not sharply inturned at the top
7 Hind limbs (**a**) longer than front limbs (**b**)
8 Shin no longer than thigh
9 Five digits per "hand" and foot

These species represent the thecodont suborders.

1 Erythrosuchus was an early proterosuchian, with a stout, squat body, thick limbs, large head, and rather short tail. Length: 4.5m (14ft 9in). Time: Early Triassic. Place: South Africa.

2 Euparkeria was a small, light pseudosuchian hunter. It rose on long hind legs to sprint, the tail balancing the head and neck. Rows of armour plates ran down the back. Length: 60cm (2ft). Time: Early Triassic. Place: South Africa.

3 Stagonolepis was a heavy-bodied aëtosaur encased in bony armour plates. It had a pig-like snout with weak teeth. Perhaps it grubbed for roots. Length: 3m (10ft). Time: Late Triassic. Place: Europe.

4 Rutiodon was a phytosaur – a crocodile-like reptile but with nostrils in a bump almost between the eyes, not at the snout tip. Such fish-eaters dominated pools and rivers until replaced by crocodiles. Length: 3m (10ft). Time: Late Triassic. Place: North America and Europe.

Crocodilians

Crocodilians are living fossils, the last surviving archosaurs. Their bulky, armoured bodies, long, deep, flattened swimmers' tails, short, sturdy limbs, and long, strong, toothy, flesh-eaters' jaws resemble those of crocodiles alive 100 million years ago.

Like dinosaurs, crocodilians evolved from thecodonts 200 million years or more ago. Fossils show they dominated pools and rivers worldwide when climates almost everywhere were warm. Small, early protosuchian crocodiles gave rise to the larger mesosuchians, some designed for life at sea. From mesosuchians came the strange, land-based sebecosuchians, and the eusuchians – the one suborder that survives today.

Where they lived
This map shows finds of the four fossil crocodiles described on these pages.
1 *Protosuchus*
2 *Metriorhynchus* (which lived in seas worldwide)
3 *Baurusuchus*
4 *Deinosuchus*

Nostrils fore or aft
Nostril position helps us to distinguish crocodilian fossils from those of phytosaurs.
a Crocodilian nostrils (an enclosed nasal passage permits breathing while eating)
b Phytosaur nostrils

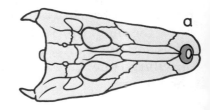

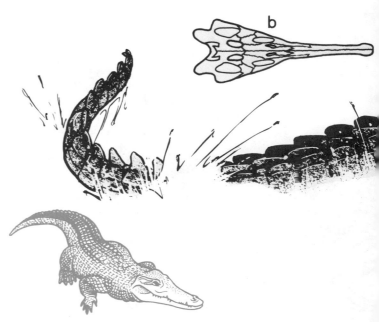

Here is one example from each suborder.

1 **Protosuchus** from Arizona was a protosuchian crocodile. It had a short, sharp-toothed skull and rather long legs. Maybe it lived mainly on land. Time: Late Triassic or Early Jurassic.

2 **Metriorhynchus** was a marine mesosuchian with limbs evolved as flippers, a tail fin, very long jaws with sharp fish-eater's teeth, and no bony armour. Length: 3m (10ft). Its Mid-Late Jurassic fossils occur in Europe and South America.

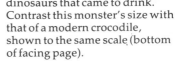

3 **Baurusuchus** was a sebecosuchian from Brazil. It had a short, deep, flattened skull, with few teeth (the front ones very large), and sideways-facing eyes. Probably it lived on land. Length: maybe 1.5m (5ft). Time: Late Cretaceous.

4 **Deinosuchus,** an eusuchian, was the largest-ever crocodile, with immense jaws. It lived in Late Cretaceous Texas and must have eaten small and medium sized dinosaurs. Length: 16m (52ft 6in).

Terror of the dinosaurs
The huge crocodile *Deinosuchus* lurked in rivers and ambushed dinosaurs that came to drink. Contrast this monster's size with that of a modern crocodile, shown to the same scale (bottom of facing page).

Pterosaurs

Pterosaurs ("winged lizards") included the first and largest flying backboned animals. They evolved from Triassic pseudosuchian gliding reptiles, lasted 130 million years, and died out at the end of the Cretaceous Period, 65 million years ago. They had skin wings stretched between the limbs and body; light, but strong, skeletons; and well-developed powers of sight and wing control. Some walked like birds, others shuffled awkwardly and roosted bat-like. Most caught fish or other prey. Two main groups evolved: first rhamphorhynchoids, with teeth and tails; then pterodactyloids, which lacked these and so shed needless weight. Large pterodactyloids were poor fliers and mostly soared or glided, where air rose over heated land or winds blew up sea cliffs.

How pterosaurs began (above) The gliding pseudosuchian *Sharovipteryx* might have given rise to pterosaurs (though these flew with their fore limbs). It lived in Early Triassic times in what is now Soviet Central Asia.

Pterosaur body plan
A pterosaur skeleton is shown here to scale with an albatross. These features adapted a pterosaur for flight.
a Skin wing membrane
b Arm bones and long fourth-finger bone support wing membrane
c Pteroid bone supports a front wing flap that helps to prevent stalling
d Clawed fingers 1–3 serve as hooks for roosting
e Light, strong skeleton with hollow, air-filled bones
f Fused vertebrae reinforce the shoulder area
g Enlarged breastbone anchors wing muscles
h Strong joint linking shoulder blade to spine and breastbone
i Large eyes with keen vision
j Brain well developed for sight and co-ordinated flight

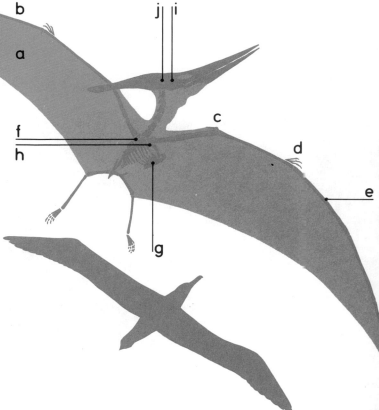

These examples show special features and trends.

1 **Dimorphodon,** an Early Jurassic pterosaur, from southern England, had a big head, biting teeth, and a long tail. Wingspan: 1.5m (5ft).

2 **Sordes** had fur to trap body heat (maybe all pterosaurs were warm-blooded). This small rhamphorhynchoid comes from Late Jurassic rocks in the USSR (Kazakhstan).

3 **Pteranodon** was a huge, tailless, toothless pterodactyloid. Its "weather vane" head crest might have kept it heading into wind. It zoomed off sea cliffs, caught fish in its beak, and stored them in a throat pouch for its young. Wingspan: 8m (26ft). Time: Late Cretaceous. Place: USA (Delaware, Kansas, and Texas) and Japan.

4 **Quetzalcoatlus,** or "feathered serpent", was the largest known pterosaur: a long-necked beast that soared on hot air and gobbled carrion. Wingspan: 11–12m (36–39ft). Weight: 86kg (190lb). Time: Late Cretaceous. Place: Texas and Alberta.

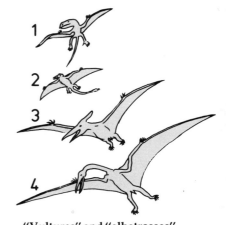

"Vultures" and "albatrosses"
Weak fliers, giant pterosaurs relied on moving air to keep them borne aloft.
A *Quetzalcoatlus* soared on thermal currents rising from land heated by the Sun.
B *Pteranodon* launched into the winds that blew up sea cliffs.

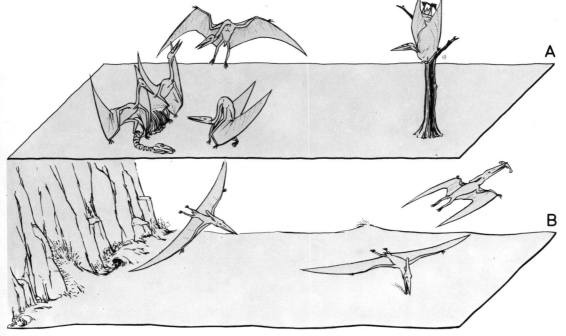

A

B

©DIAGRAM

Saurischian dinosaurs 1

Dinosaurs ("terrible lizards") were probably the most successful-ever backboned animals to live on land. They evolved from pseudosuchian thecodonts some 205 million years ago and dominated lands worldwide for 140 million years – all through the rest of Mesozoic time. Their more than 350 genera included bipeds and quadrupeds: flesh-eaters and plant-eaters. Some were the largest-ever land animals. A few grew no bigger than a chicken. Many were probably warm-blooded. Almost all walked and ran on the toes, with legs straight down below the body, like a horse or an ostrich.

Two orders of these archosaurs evolved. Saurischians (the "lizard-hipped") comprised two suborders. One, the theropods or "beast feet", contained the flesh-eating dinosaurs. These bipeds ranged from small lizard-catchers to monsters as heavy as an elephant. A few small theropods were brainier than any reptile now alive.

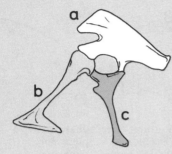

Saurischian hip bones (above)
A saurischian's hip girdle had a forward-pointing pubis, as in most reptiles.
a Ilium **b** Pubis **c** Ischium

Theropod skulls (right)
Shown here are four theropod skulls of contrasting shapes and sizes. A human skull has also been included for scale.
A *Tyrannosaurus*, one of the largest of all flesh-eating dinosaurs, had big, sharp teeth shaped like serrated "steak-knife" blades.
B *Allosaurus*, a large Jurassic scavenger or big-game hunter, had sabre-like teeth designed for piercing flesh.
C *Ornithomimus*, a toothless "ostrich dinosaur", probably ate insects, lizards, leaves, fruit, and seeds.
D *Compsognathus*, no bigger than a chicken, had small, sharp teeth and ate lizards.
E Human skull, to scale.

©DIAGRAM

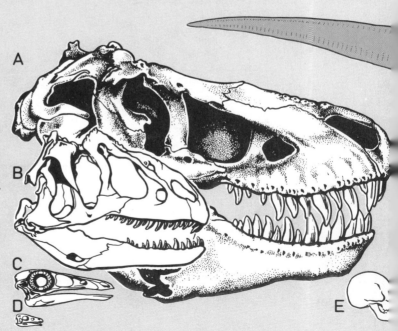

Some experts divide theropods into three infraorders. Here is one example of each.

1 **Coelurus** belonged to the coelurosaurs – small, light, sprinters with long legs, a long tail, and sharp claws and teeth (but some had toothless beaks). Length: 2m (6ft 6in). Time: Late Jurassic. Place: Wyoming, USA.

2 **Deinonychus** was a deinonychosaur – a small, fierce hunter that could stand on one leg, balanced by its stiffened tail, and slash out with a deadly "switchblade" toe claw. Length: 2.4–4m (8–13ft). Time: Early Cretaceous. Place: western USA.

3 **Tyrannosaurus** was one of the largest carnosaurs (great flesh-eating dinosaurs). It had a mighty head and body, huge legs and toe claws, savage fangs, and massive jaws. Arms were small but muscular. It killed big plant-eating dinosaurs or ate ones already dead. Length: 12m (39ft). Height: 5.6m (18ft 4in). Weight: 6.4 tonnes. Time: Late Cretaceous. Place: North America and China.

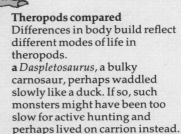

Theropods compared
Differences in body build reflect different modes of life in theropods.
a *Daspletosaurus*, a bulky carnosaur, perhaps waddled slowly like a duck. If so, such monsters might have been too slow for active hunting and perhaps lived on carrion instead.
b *Dromiceiomimus* was an ostrich dinosaur, capable of running faster than a horse.
c *Stenonychosaurus* was an agile sprinter with a brain larger than an emu's.
d Man, shown to scale.

Saurischian dinosaurs 2

The second saurischian suborder, the sauropodomorphs ("lizard-feet forms"), included the largest dinosaurs of all. Pioneers were the prosauropods: two-legged and four-legged dinosaurs, some lighter than a man. Their mostly Triassic infraorder included perhaps the first plant-eating dinosaurs. In the Jurassic Period these gave way to the colossal, four-legged sauropods – the largest-ever land animals. These giants raised long necks to browse on trees fringing warm, sluggish, lowland rivers. Their vast bulk made most of them safe from all enemies except carnosaurs and giant crocodiles. Sauropods endured as long as any dinosaurs, but their heyday was in Late Jurassic times. Rich fossil finds have come from China, the United States, and Tanzania.

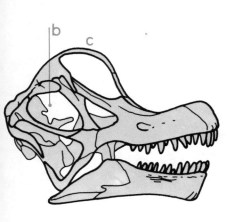

Sauropod features
The *Brachiosaurus* skull (above) and skeleton with outlined body (right) include features typical of many sauropods.
a Relatively small head
b Very small brain (bulk for bulk the smallest brain of any vertebrate)
c Some skull bones reduced to struts to save weight
d Long, flexible neck
e Spinal bones hollowed out to reduce weight
f Vast, heavy body

g Massive limbs, with thick, heavy bones
h Short, strong hind feet; toes 1–3 had long claws
i Each thumb had one long claw
j Long tail

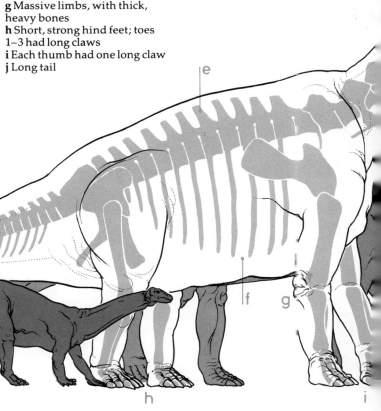

1

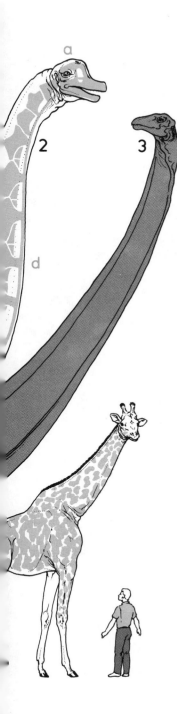

1 **Plateosaurus,** a big prosauropod, had a bulky body, rather long neck, small head, long tail, big, strong hind limbs, and shorter forelimbs. The five-fingered "hand" had a great thumb claw. This beast ate plants and maybe meat, and lived in Late Triassic Western Europe. Length: 8m (26ft).

2 **Brachiosaurus,** a brachiosaurid sauropod, was one of the most massive dinosaurs of all. Length: about 23m (75ft) or more. Height: 12m (39ft). Weight: 78 tonnes or more. Time: Late Jurassic. Place: USA, Portugal, Algeria, and Tanzania.

3 **Diplodocus** belonged to the diplodocids – a family of lightweight sauropods (some perhaps the longest dinosaurs of all) with snaky necks and whiplash tails. Length: 26.6m (87ft). Weight: 10.6 tonnes. Time: Late Jurassic. Place: western USA.

Sauropodomorph sizes (left)
Three sauropods are here shown to the same scale as a giraffe and a man.
1 *Plateosaurus*
2 *Brachiosaurus*
3 *Diplodocus*

Where they lived (below)
This map shows sites of fossil finds of the dinosaurs shown on these two pages.
1 *Plateosaurus*
2 *Brachiosaurus*
3 *Diplodocus*

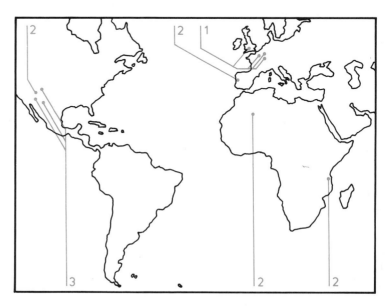

©DIAGRAM

Ornithischian dinosaurs 1

Ornithischian ("bird hipped") dinosaurs, the second dinosaur order, get their name from hip bones designed like those of birds. Almost all were herbivores, with jaws superbly engineered for cropping and chewing leaves. Maybe this is why ornithischian species in time outnumbered those of sauropods, whose teeth were less effective. Saurischians or perhaps thecodonts gave rise to ornithischians in Late Triassic or Early Jurassic times. We shall look briefly at their four suborders, first the ornithopods.

Ornithopods ("bird feet") comprised dozens of species that walked or ran on long hind limbs. Small, early, agile sprinters no larger than a big dog gave rise to beasts as heavy as elephants. Here are examples from four families.

1 **Hypsilophodon** was a dinosaur "gazelle": a small, lightweight sprinter with long shins and feet, short arms, and a long stiffened tail to balance the head and body as it ran. It had ridged, self-sharpening teeth. Length: 1.8m (6ft). Time: Early Cretaceous. Place: England, Spain, Portugal, and USA.

Ornithischian hip bones
In ornithischian dinosaurs the pubis pointed backward.
a Ilium b Pubis c Ischium

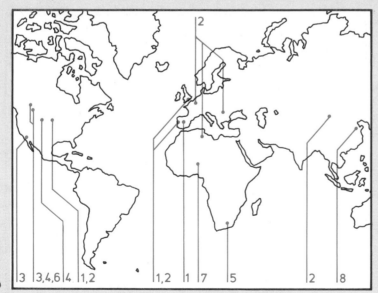

Where they lived
This map shows fossil finds of ornithopods described or pictured on these two pages. Items 1–4 tally with those so numbered in the text.
1 *Hypsilophodon*
2 *Iguanodon*
3 *Lambeosaurus*
4 *Pachycephalosaurus*
5 *Fabrosaurus*
6 *Stegoceras*
7 *Ouranosaurus*
8 *Shantungosaurus*

©DIAGRAM

2 **Iguanodon** was big and bulky with relatively large arms, a toothless beak, and hoofed nails. It roamed swampy lowlands, mostly on all fours. Length: up to 9m (29ft 6in). Weight: up to 4.5 tonnes. Time: Early Cretaceous. Place: northern continents.

3 **Lambeosaurus** belonged to the hadrosaurids or duckbilled ornithopods – Late Cretaceous beasts with wide, toothless beaks but up to 2000 cheek teeth: more than any other dinosaurs. Some hadrosaurids had head crests or "nose flaps" which they could blow up like balloons. Some grew far larger than their iguanodontid ancestors. Duckbills roamed all northern continents. *Lambeosaurus* from western North America grew up to 15m (49ft).

4 **Pachycephalosaurus,** a pachycephalosaurid ("bone-headed lizard") had a thick crash-helmet skull. Maybe rival males banged heads and winners ruled herds of females. Length: 4.6m (15ft). Time: Late Cretaceous. Place: North America. (Most other boneheads came from China or Mongolia.)

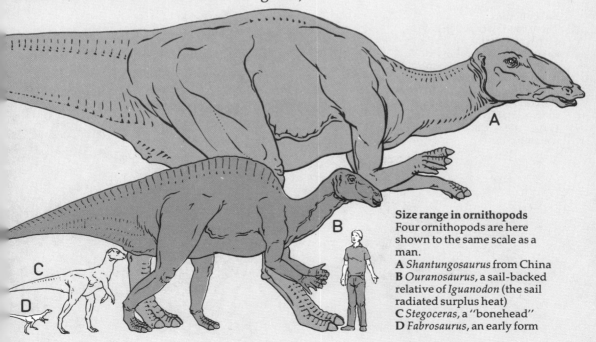

Size range in ornithopods
Four ornithopods are here shown to the same scale as a man.
A *Shantungosaurus* from China
B *Ouranosaurus*, a sail-backed relative of *Iguanodon* (the sail radiated surplus heat)
C *Stegoceras*, a "bonehead"
D *Fabrosaurus*, an early form

127

Ornithischian dinosaurs 2

Our brief look at the dinosaurs ends with three suborders of mostly four-legged ornithischians: the plated, armoured, and horned dinosaurs. Like their ornithopod relatives, all three had horny beaks for cropping plants. But unlike ornithopods most were not designed to run fast from danger. Instead they relied seemingly on body armour to protect them from the carnosaurs.

Plated dinosaurs (stegosaurs) had thick skin with bony plates or spikes that jutted from the back and tail. Armoured dinosaurs (ankylosaurs) had back and flanks encased in flexible armour made up of bony chunks and spikes covered with horny sheaths. Horned dinosaurs (ceratopsians) had a backswept bony crest protecting neck and shoulders; many sprouted rhinoceros-like horns. Stegosaurs evolved first, ankylosaurs second, and ceratopsians third. But ankylosaurs might have sprung from *Scelidosaurus*, an Early Jurassic dinosaur pre-dating even stegosaurs.

Bulky bodies
Three ornithischian dinosaurs, each representing a different suborder, are here shown to scale with a light battle tank.
A *Stegosaurus*, a plated dinosaur (stegosaur)
B *Ankylosaurus*, an armoured dinosaur (ankylosaur)
C *Triceratops*, a horned dinosaur (ceratopsian)

1 **Scelidosaurus,** perhaps a proto-ankylosaur, had seven rows of bony studs and spikes set in its back. Length: 3.5m (11ft 6in). Time: Early Jurassic. Place: England and Tibet.

2 **Stegosaurus,** the largest stegosaur, had two rows of plates (some huge) along the neck and back, and four spikes to guard the tail. Length: up to 9m (30ft). Time: Late Jurassic. Place: USA.

3 **Ankylosaurus,** the largest ankylosaur, had a thickened skull and crosswise bands of bony plates and studs protecting back and tail. The tail ended in a bony club. Length: up to 10.7m (35ft). Time: Late Cretaceous. Place: western North America.

4 **Triceratops,** among the last and largest of the ceratopsians, reached 9m (30ft) and 5.4 tonnes. It had a short neck frill, short nose horn, and two long brow horns. Time: Late Cretaceous. Place: western North America.

©DIAGRAM

Mammal-like reptiles 1

Mammal-like reptiles form the subclass Synapsida – reptiles with one hole low in each side of the skull behind the eye. All were quadrupeds. Early kinds were sprawlers like their cotylosaur ancestors. But evolution produced species that stood more erect, had body hair and several kinds of teeth, and were warm-blooded – features found in their better-known descendants, mammals.

Pelycosaurs or "basin-shaped (pelvis) lizards" were the earlier, more primitive, of two synapsid orders to appear. They arose by 280 million years ago. In Early Permian times, big, sprawling pelycosaur flesh-eaters and herbivores dominated life on land, at least in North America and Europe, where almost all their fossils have been found. They died out about 250 million years ago.

Synapsid family tree
This shows orders (**1–3**) and suborders (**a–f**).
1 Cotylosaur ancestors
2 Pelycosauria (primitive mammal-like reptiles)
a Edaphosauria (herbivores)
b Sphenacodontia (carnivores)
c Ophiacodontia (early forms)
3 Therapsida (advanced mammal-like reptiles)
d Phthinosuchia (ancestral therapsids)
e Theriodontia (advanced flesh-eaters)
f Anomodontia (herbivores)

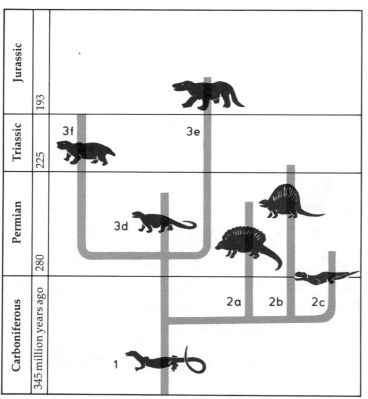

Two types of jaws (right)
Here we compare the skulls of flesh-eating *Dimetrodon* and plant-eating *Edaphosaurus*.
A *Dimetrodon*
a Long, deep, narrow skull
b Strong jaws with wide gape
c Small chewing teeth
d Two pairs of upper canine teeth: saw-edged blades
e "Step" in upper jaw
f Biting and grasping "incisors"
B *Edaphosaurus*
a Short, fairly shallow skull
b Straight-edged jaws (no step)
c Blunt teeth, all more or less alike
d Toothplates in mouth roof – *Edaphosaurus* could have crushed tough seed-fern leaves and maybe even mollusc shells
Sails as radiators (far right)
Long spines jutting from the backbone supported *Dimetrodon's* skin sail. If the pelycosaur stood sideways to the Sun this heated the sail's blood supply, so warming the whole body. If *Dimetrodon* faced away from the Sun, its sail shed heat, and its body cooled.

These examples show features of the pelycosaurs' three suborders: ophiacodonts, sphenacodonts, and edaphosaurs.

1 **Ophiacodon,** an ophiacodont, had a low-slung, lizard-like body, long hind legs, narrow, deep, long head, and jaws equipped with many sharp teeth. Probably it hunted fish in rivers. Length: 3.7m (12ft). Time: Early Permian. Place: Texas, USA.

2 **Dimetrodon** was a sphenacodont – a big flesh-eater with long, sharp "steak-knife" teeth and powerful jaws. It was one of the first backboned land animals able to kill beasts its own size. A huge skin "sail" rose from its back. Length: 3.5m (11ft 6in). Time: Early–Mid Permian. Place: Texas and Oklahoma, USA.

3 **Edaphosaurus,** an edaphosaur, was a large herbivore with blunt teeth, some in the mouth roof. A long, high skin "sail" ran down its back. Length: 3.3m (11ft). Time: Late Pennsylvanian–Early Permian. Place: USA and Europe.

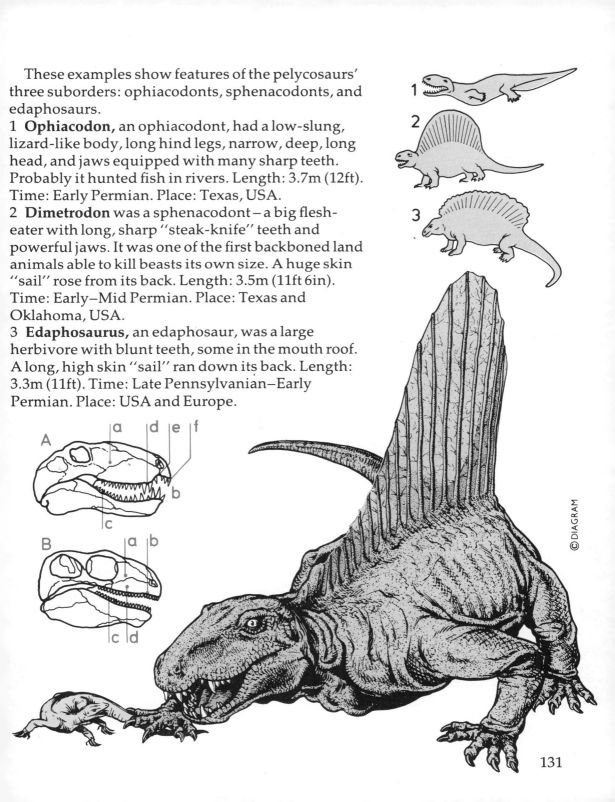

©DIAGRAM

131

Mammal-like reptiles 2

In time an advanced, varied order of mammal-like reptiles, the therapsids (literally "mammal arch"), took over from their ancestors the pelycosaurs. Therapsids flourished from Mid Permian to Early Jurassic times. In the Late Permian they formed the chief flesh-eaters and plant-eaters living on dry land. Scientists have found Permian therapsid fossils worldwide, especially in South Africa and the USSR.

The therapsids comprised three suborders. Here we give examples of two: the primitive, ancestral phthinosuchians, and their "dead-end" offshoot the anomodonts. The latter were small to large plant-eaters and flesh-eaters in four infraorders: dromasaurs, dinocephalians ("terrible heads"), venyukoviamorphs, and dicynodonts ("two dog-like teeth").

1 **Phthinosuchus** belonged to the phthinosuchians – "old-fashioned" therapsids but with a large skull opening behind each eye, one pair of canine teeth per jaw, and a more upright stance than the pelycosaurs. Length: 1.5m (5ft). Time: Mid Permian. Place: USSR.

2 **Galepus** belonged to the dromasaurs – little, lightweight insect-eaters with jaws hinged well below the tooth row. Length: maybe 30cm (1ft). Time: Mid Permian. Place: South Africa.

3 **Moschops** was a plant-eating dinocephalian. It had a thick, short, dome-shaped skull, peg-like cropping teeth, squat, heavy body, sloping back, and short tail. Stocky limbs held its body well off the ground. Big flesh-eating dinocephalians might have attacked it. Length: 2.4m (8ft). Time: Mid Permian. Place: South Africa.

4 **Venyukovia** belonged to the venyukoviamorphs –
big, partly beaked herbivores with a deep lower jaw,
mostly short teeth, and a few big, stubby front teeth.
Time: Mid Permian. Place: USSR.

5 **Lystrosaurus** belonged to the dicynodonts –
abundant, worldwide plant-eaters. They had a short,
broad body, short tail, strong legs, big holes in the
skull behind the eyes, a horny beak, and a toothless
mouth or just two upper tusks. *Lystrosaurus* was an
Early Triassic reptilian "hippo", at home by lakes
and rivers. Length: 1m (3ft 3in). Place: Antarctica,
South Africa, India, and China.

Hunters and hunted
South Africa's Karroo rock beds
hold fossil bones of big
dinocephalians like *Moschops*
(**A**) and *Titanosuchus* (**B**).
Moschops peaceably ate plants,
but *Titanosuchus*'s long, heavy
jaws had sharp incisors and
long, stabbing canine teeth, for
tackling big game.

A

B

©DIAGRAM

133

Mammal-like reptiles 3

The most mammal-like of all mammal-like reptiles belonged to the therapsids' third suborder: the theriodonts ("mammal toothed"). These flesh-eaters were mostly small to medium-sized, with teeth and many bones designed astonishingly like a mammal's. Some were certainly warm-blooded and had a covering of body hair. A few perhaps even suckled young. Only the jaw and hearing mechanism marks these off from mammals, their direct descendants.

Theriodonts flourished about 250–170 million years ago. They spread worldwide, but South Africa's Permian rocks are the richest source of fossils. Here are examples from the theriodonts' six infraorders (some known well only from skulls). All but example 5 are from South Africa.

1 **Lycaenops** belonged to the gorgonopsians – plentiful Permian flesh-eaters derived from pelycosaur forebears. They had rather low-slung bodies, heavy skeletons and "sabre" teeth. Length 1m (3ft 3in). Time: Late Permian.

2 **Lycosuchus** represents the therocephalians. These had as few toe and finger bones as mammals, a skull crest, and a large hole in the skull behind each eye. Some were powerfully-built carnivores. Length: 1.8m (6ft). Time: Mid Permian.

3 **Bauria** was one of the bauriamorphs, with mammal-like teeth and skull but old-fashioned lower jaw. *Bauria* seemingly had broad, grinding back teeth inset from the jaw rim, so perhaps space for cheeks storing half-chewed food. Length: maybe 1m (3ft 3in). Time: Early Triassic.

4 **Cynognathus** was a wolf-sized cynodont predator or scavenger, with dog-like skull and teeth, and limbs held fairly well below the body, a help in running fast. Length: 1.5m (5ft). Time: Early–Mid Triassic.

5 **Oligokyphus** belonged to the tritylodonts: small, rodent-like last survivors of all mammal-like reptiles. Length: 50cm (20in). Time: Late Triassic–Early Jurassic. Place: England and Portugal.

6 **Diarthrognathus** was an ictidosaur ("weasel lizard"), in the group arguably ancestral to the mammals. Its jaw hinged almost like a mammal's. Length: maybe 40cm (16in). Time: Late Triassic.

Cynognathus skull
This reveals a tendency towards mammalian features.
A Mouth and nasal passage separated, allowing breathing while eating
B Differentiated teeth: incisors, canines and ridged cheek teeth for fast chewing (speeding digestion to provide a high energy output)

Inner jaws
The inner jaws of *Cynognathus* and two early mammals (not to scale) illustrate a trend to fewer bones. The dentary is here shown tinted.
a *Cynognathus* jaw: seven bones, the dentary relatively larger than in early synapsids
b *Morganucodon* jaw: four bones, the dentary by far the largest
c *Spalacotherium* jaw: one bone, the dentary

Mammal-like features
This whole-body restoration of *Cynognathus* is based on key features found in skeletal remains.
1 Body covering of hair inferred from whisker pits in the snout
2 Mammal-like posture (knees and elbows held beneath the body) inferred from bones of limbs, hips and shoulders

Chapter 7   FOSSIL BIRDS

Aves (birds) is the most recently evolved of all nine classes of backboned animals. We can divide birds into three subclasses: Archaeornithes, comprising only *Archaeopteryx*, the first known bird; Odontoholcae, the toothed Cretaceous birds; and Neornithes, the rest. This short chapter gives prehistoric examples from the birds' more than 30 orders, most of which survive today.

A nineteenth-century book illustration shows a flock of moa skeletons in New Zealand's Canterbury Museum. Unlike most prehistoric birds, these recently extinct flightless giants left behind a wealth of bones. They enabled scientists to build up detailed reconstructions. (The Mansell Collection.)

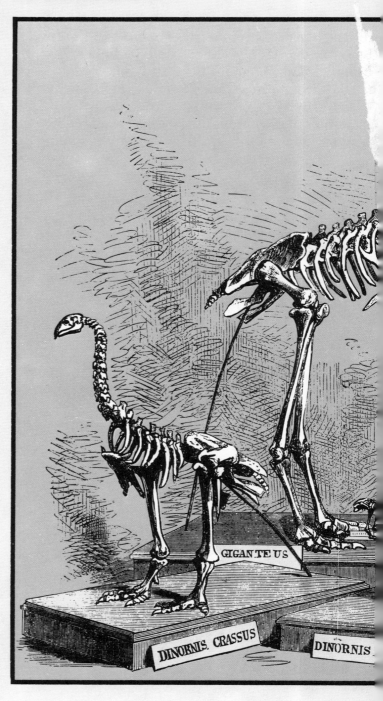

GIGANTEUS

DINORNIS. CRASSUS

DINORNIS

ELEPHANTIPUS

ROBUSTUS

DINORNIS CASUARINUS

137

About birds

Birds are warm-blooded, backboned animals with covering of feathers. Most have front limbs evolve as wings capable of flapping, soaring, or gliding flight. Early birds had teeth and a long bony tail co Later birds evolved a more lightweight structure: with a toothless beak, a tail of lightweight feathers only, and hollow, air-filled bones for buoyancy. Flying birds gained a powerful breastbone to help anchor big, strong wing muscles. The first birds evolved about 140 million years ago. Some experts think birds' ancestors were small carnivorous coelurosaurs: dinosaurs like little *Compsognathus*. Other experts believe birds evolved directly from the dinosaurs' own ancestors, pseudosuchian thecodonts like *Euparkeria*. At first birds shared the air with pterosaurs. But birds have long outlived these creatures. This might be partly because birds'

Euparkeria
Euparkeria walked on all fours but could rise on its long hind legs to chase prey or escape from enemies. Early small pseudosuchian thecodonts like this gave rise to crocodiles, pterosaurs, dinosaurs, and maybe birds. But a long time gap separates the last known pseudosuchian from the first known bird.

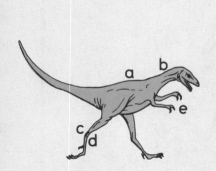

Compsognathus
Tiny coelurosaur dinosaurs similar to *Compsognathus* might have given rise to the first known birds. Both shared over 20 anatomical similarities. Five are shown here.
a Short body
b Slim, flexible neck
c Very long legs
d Stiff ankle joints
e Long fingers

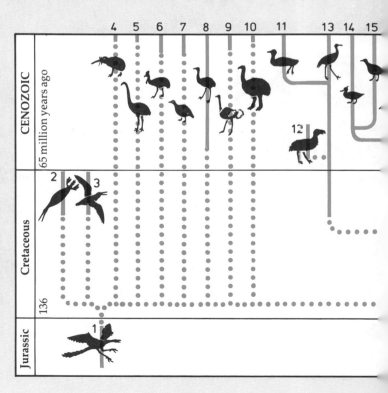

feathered wings survive injuries better than the pterosaurs' more fragile skin wings.

The class Aves (birds) contains some 30 orders, most with living representatives. No one knows exactly how and when each order began, or how all orders are related. This is largely because birds' fragile bones are seldom fossilized. Indeed, some species are known from little more than the fossil impression of a feather, or from a fossil footprint. But certain rocks have produced key fossil bird remains: for instance fine-grained limestone from Solnhofen in southern West Germany, the Niobrara Chalk of Kansas, and mudstone rocks of Utah and Wyoming. Finds in places such as these tell us that a number of orders of modern birds – and maybe even a few modern families – have flourished 50 million years or more.

Family tree of birds
Zoologists divide the class Aves (birds) into three subclasses, containing three superorders, subdivided into the 33 orders numbered below.
A Subclass Archaeornithes
1 Archaeopterygiformes (*Archaeopteryx*)
B Subclass Odontoholcae (Odontognathous or toothed Cretaceous birds)
2 Hesperornithiformes
3 Ichthyornithiformes
C Subclass Neornithes: Palaeognathous birds
4 Apterygiformes (kiwis)
5 Dinornithiformes (moas)
6 Casuariformes (emus etc)
7 Tinamiformes (tinamous)
8 Rheiformes (rheas)
9 Struthioniformes (ostriches)
10 Aepyornithiformes (elephant birds)
C Subclass Neornithes: Neognathous birds
11 Podicipediformes (grebes)
12 Diatrymiformes (*Diatryma*)
13 Gruiformes (cranes, rails)
14 Anseriformes (ducks)
15 Galliformes (grouse etc)
16 Charadriiformes (shorebirds)
17 Procellariiformes (albatrosses etc)
18 Sphenisciformes (penguins)
19 Pelecaniformes (pelicans)
20 Gaviiformes (divers)
21 Ciconiiformes (storks)
22 Columbiformes (pigeons)
23 Psittaciformes (parrots)
24 Strigiformes (owls)
25 Falconiformes (hawks etc)
26 Caprimulgiformes (nightjars)
27 Apodiformes (swifts)
28 Cuculiformes (cuckoos)
29 Coliiformes (mousebirds)
30 Passeriformes (perching birds)
31 Coraciiformes (rollers)
32 Piciformes (woodpeckers)
33 Alcediniformes (kingfishers)

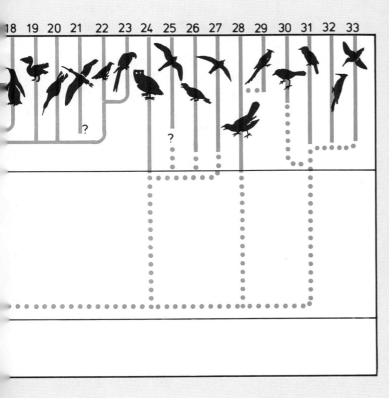

18 19 20 21 22 23 24 25 26 27 28 29 30 31 32 33

©DIAGRAM

Early birds

Birds shown here represent three early groups with teeth. *Archaeopteryx*, the first known bird, resembled a small coelurosaurian dinosaur but had feathers and long "arms". Wings, wishbone, and the angled bones of its shoulder girdle adapted *Archaeopteryx* for flight. Some experts think it climbed trees and fluttered down. Others believe it took off by sprinting into a headwind after insects. *Hesperornis* and its kin probably hunted fish just below the sea surface, and nested on lonely coasts or offshore islands. *Ichthyornis* may have flown above the sea and plunged to seize small fish, as terns do now. It might have given rise to modern shore birds.

1 **Archaeopteryx** ("ancient wing") had feathered wings but also unbird-like features: small teeth in the jaws, three-clawed fingers jutting from each wing, and a tail with a long, thin, bony core. Length: about 1m (about 3ft). Time: Late Jurassic. Place: Bavaria, southern West Germany. Order: Archaeopterygiformes.

2 **Hesperornis** ("western bird") resembled a large diver. It had a long, slim, pointed beak rimmed with teeth, and vestigial wings. It swam by thrusting water back with big lobed feet. Legs joined the body far back, so *Hesperornis* shuffled clumsily on land. Length: up to 1.5m (5ft). Time: Late Cretaceous. Place: North America. Order: Hesperornithiformes (toothed divers).

3 **Ichthyornis** ("fish bird") was a small, stout, tern-like bird with long, pointed wings, small feet and a long, slim beak armed with small curved teeth. *Ichthyornis* is the earliest known bird with a keeled breastbone to help support flight muscles. Height: 20cm (8in). Time: Late Cretaceous. Place: North America (eg Kansas, Texas, Alabama). Order: Ichthyornithiformes (toothed tern-like birds).

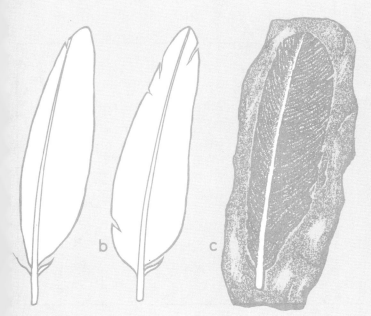

The first known bird (above)
Archaeopteryx might have used
its clawed limbs for climbing
trees, then flapped and fluttered
down. Ability to climb and fly
would have helped it to escape
enemies and capture agile prey.

Feathers and flight (left)
Feathers help to prove that
Archaeopteryx flew.
a Flying bird's wing feather: its
asymmetrical design helps it act
as an aerofoil.
b Flightless bird's wing feather:
its symmetrical design is useless
for flight.
c *Archaeopteryx* wing feather: its
design is asymmetrical.

© DIAGRAM

141

Flightless birds

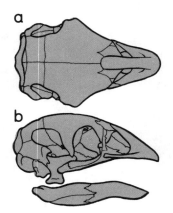

A moa's skull (above)
These views show a moa skull seen from above (**a**) and from the right (**b**). The beak was shaped for cropping plants, not rending flesh.

Where they lived (below)
This map depicts homes of the four groups of extinct flightless birds represented by those named on the facing page.
1 Elephant birds
2 Terror cranes
3 Moas
4 Phorusrhacids

Most dead land birds make poor fossils: their fragile skeletons are soon eaten or just rot away. But big, flightless running birds have left their remains in Cenozoic rocks. Many of these birds are collectively called "ratites" from the Latin *ratis* ("raft"), for the breastbone was flat and raft-like, not keeled to anchor powerful flight muscles like the breastbone of a flying bird. Some experts think all prehistoric ratites and their living descendants the emu and ostrich shared one ancestor; others argue that different ratites had separate beginnings.

The flightless birds shown here were huge. Some filled the roles of cattle, others behaved like big cats. Such monsters tended to appear in lands with no large mammal herbivores or carnivores. Early ratites left only fossils, but *Dinornis* and *Aepyornis* survive as actual bones, found preserved in swamps. Archaeologists have even identified some moas' stomach contents. There are also *Aepyornis maximus* eggs – the largest known eggs to have been laid by any bird.

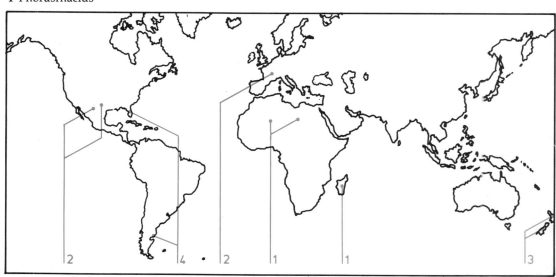

1 **Aepyornis maximus** ("greatest of the high birds") was possibly the heaviest-ever bird at 440kg (970lb). A Madagascan elephant bird, it was the "roc" of legend, extinct by AD1700.

2 **Diatryma steini,** a "terror crane", stood 2m (6ft 6in) high and probably killed prey with its huge clawed feet and massive "parrot's" beak. It lived in North America 50 million years ago.

3 **Dinornis giganteus** (" giant terrible bird") was the tallest known bird, 3.5m (11ft 6in) high. This New Zealand giant moa was a browser that may have died out only about 400 years ago.

4 **Phorusrhacos longissimus** stood 2m (6ft 6in) tall and probably devoured goat-sized creatures with its huge "eagle's" beak. This savage hunter prowled Patagonia about 20 million years ago.

Grounded giants
Four giant flightless birds appear here on the same scale as a chicken.
1 *Aepyornis maximus*
2 *Diatryma steini*
3 *Dinornis giganteus*
4 *Phorusrhacos longissimus*

143

Water birds

Birds shown here are fossil or living examples of the eight orders of water birds and marsh birds. All except Ciconiiformes were probably derived from shorebirds.

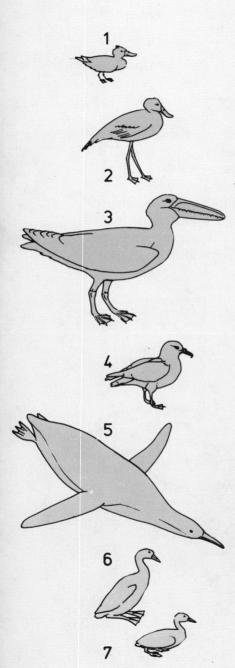

1 **Stictonetta**, Australia's freckled duck is a primitive living member of the order Anseriformes (ducks, geese, swans), an order dating from the Eocene.

2 **Presbyornis** was a long-legged wader with a mainly duck-like skull. It might have given rise to ducks, geese, and swans. Vast flocks bred and fed on algae in salty, shallow lakes. Time: Early Eocene. Place: North America, and maybe worldwide. Order: Charadriiformes – shorebirds including gulls, terns, auks, and many waders.

3 **Osteodontornis** was among the largest-ever flying birds. Maybe it seized squid from the sea surface in its long bill rimmed with tooth-like bony spikes. Wingspan: up to 5.2m (17ft). Time: Miocene. Place: California. Order: Pelecaniformes – web-footed, fish-eating seabirds including cormorants, gannets, and pelicans.

4 **Puffinus,** an early shearwater, lived in Eocene Europe and Miocene North America. Order: Procellariiformes (albatrosses, petrels, shearwaters) – oceanic birds with long, slim, strong wings, webbed feet, and tube-shaped nostrils.

5 **Pachydyptes,** was a giant penguin about 1.6m (5ft 3in) high. Time: Eocene. Place: New Zealand. Order: Sphenisciformes (penguins) – flightless birds with wings evolved as swimming flippers. Penguins possibly evolved from petrels.

6 **Colymboides** was a diver (loon) no bigger than a small duck. Time: Eocene. Place: England. Order: Gaviiformes (divers) – swimming birds with webbed feet on legs set far back; strong fliers.

7 **Podiceps** was an early grebe. Time: Oligocene. Place: Oregon, USA. Order: Podicipediformes (grebes) – swimming birds resembling divers but with lobed not webbed feet. They probably evolved from gruiformes (cranes, rails, etc).

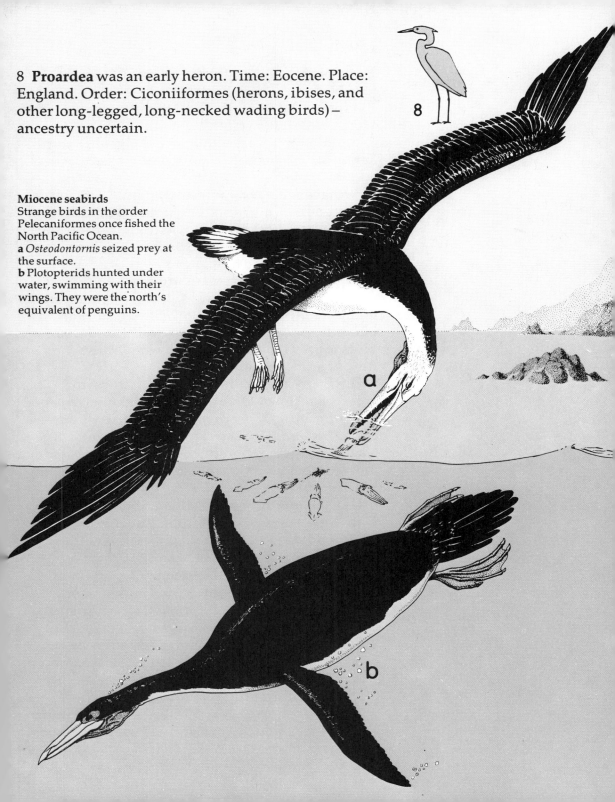

8 **Proardea** was an early heron. Time: Eocene. Place: England. Order: Ciconiiformes (herons, ibises, and other long-legged, long-necked wading birds) – ancestry uncertain.

8

Miocene seabirds
Strange birds in the order Pelecaniformes once fished the North Pacific Ocean.
a *Osteodontornis* seized prey at the surface.
b Plotopterids hunted under water, swimming with their wings. They were the north's equivalent of penguins.

a

b

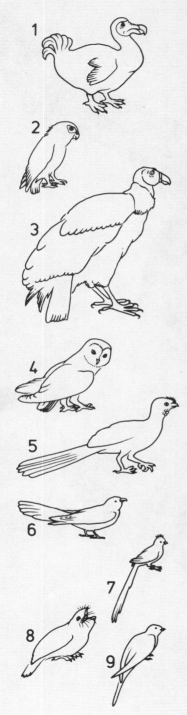

Land birds

These pages give early examples from 13 orders of land birds, including birds of prey.

1 Raphus, the dodo, was a flightless pigeon as big as a large turkey. It died out a mere three centuries ago. Order: Columbiformes (doves and pigeons) – birds derived from shorebirds as early as the Eocene epoch.

2 Archaeopsittacus was an early parrot. Time: Late Oligocene. Place: France. Order: Psittaciformes (parrots) – tropical birds with a strong, hooked bill and two toes per foot turned back for perching. They might have evolved from pigeons.

3 Argentavis, a gigantic vulture-like bird of prey, was the largest-known bird able to fly. Wingspan: up to 7.6m (25ft). Weight: 120kg (265lb). Time: Early Pliocene. Place: Argentina. Order: Falconiformes – birds of prey that mostly hunt by day. These date from Late Eocene times.

4 Ogygoptynx, the first known owl, dates from Palaeocene times. Order: Strigiformes (owls) – nocturnal birds of prey unrelated to falconiformes.

5 Gallinuloides was an early member of the Galliformes (chickens and their kin). These possibly derived from ducks and geese. Time: Eocene. Place: Wyoming, USA.

6 Dynamopterus was an early cuckoo. Time: Oligocene. Place: France. Order: Cuculiformes (cuckoos, hoatzins, touracos) – primitive land birds, most with an outer toe turned back.

7 Colius has reversible hind and outer toes, and swings acrobatically from twigs. Time: Recent. Place: Africa. Order: Coliiformes (mousebirds), dating from Miocene times.

8 Caprimulgus has long, slim wings and a short, broad bill. It hunts at twilight. Time: Pleistocene on. Place: worldwide. Order: Caprimulgiformes (nightjars, oilbirds, frogmouth, etc).

9 Aegialornis, an early swift-like bird, had scimitar-shaped wings and flew fast to catch insects. Time:

Oligocene. Place: France. Order: Apodiformes (swifts and maybe hummingbirds).

10 **Geranopterus** was an early roller from Oligocene France. Order: Coraciiformes (rollers, hornbills, hoopoes) – colourful hole-nesters with the three front toes partly joined. These were the main land birds of the Oligocene epoch.

11 **Archaeotrogon** was an early trogon from Eocene France. Order: Alcediniformes (bee-eaters, kingfishers, motmots, todies, trogons) – colourful hole-nesters resembling coraciiforms but with a unique kind of middle ear bone.

12 **Neanis,** a primobucconid, lived in Eocene Wyoming. Order: Piciformes (barbets, toucans, woodpeckers, etc) – perching birds with a reversed outer toe.

13 **Lanius,** the shrike genus, goes back to Miocene France. Order: Passeriformes ("true" perching birds) – birds with the first toe turned back. They include the songbirds and account for three-fifths of all living bird species.

©DIAGRAM

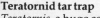

Teratornid tar trap
Teratornis, a huge condor-like vulture, here plucks flesh from a mammoth fallen in a tar pool. Feasting teratornids also tumbled in, became stuck and died. In fact the tar pools of Los Angeles are rich in Late Pleistocene bird bones. Palaeontologists have recovered those of condors, eagles, hawks, owls, ducks, geese, herons, storks, cranes, pigeons, ravens, turkeys and many perching birds.

147

Chapter 8

FOSSIL MAMMALS

Members of the class Mammalia have dominated life on land for the last 65 million years. But mammal origins go back almost three times as far as that. Starting with the first primitive mammals, this chapter covers the mammals' four subclasses: the extinct Eotheria and Allotheria; the egg-laying Prototheria with few surviving species; and the pouched and placental mammals, collectively called Theria.

Most pages deal with the placental mammals, by far the largest mammal group. We look at prehistoric members of their nearly 30 orders, 13 of which are now extinct.

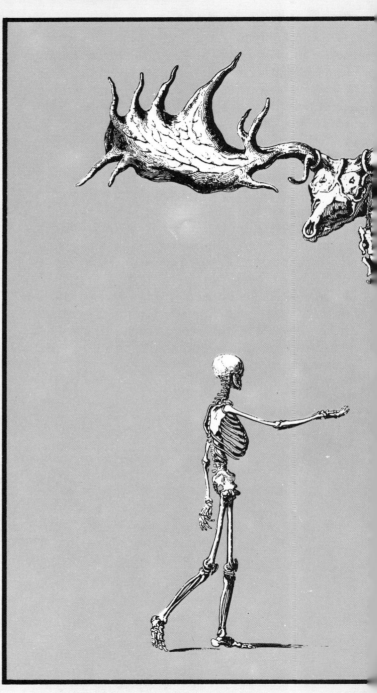

An old engraving contrasts skeletons of modern man and *Megaloceros*, the giant "Irish elk" from Pleistocene Eurasia. Dissimilar in form and size, man and deer remind us of the great diversity of mammals that evolved from tiny, shrew-like pioneers. (The Mansell Collection.)

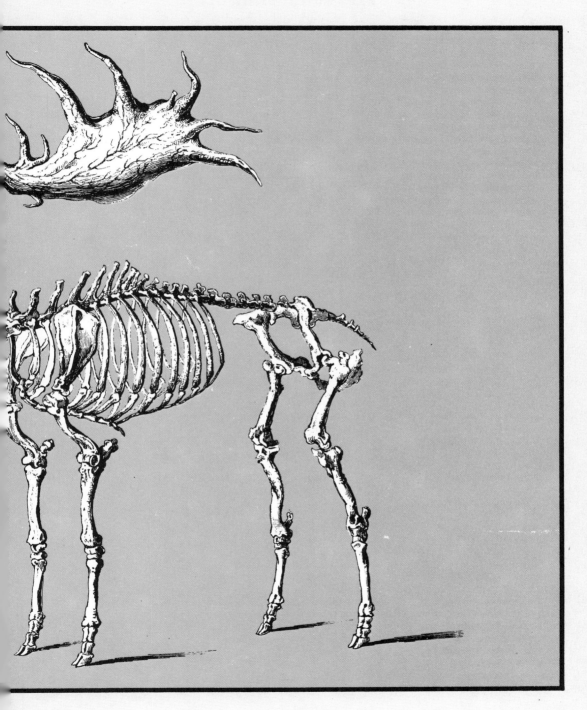

About mammals

By 190 million years or so ago, mammal-like reptiles had given rise to mammals: warm-blooded backboned animals with hair, an efficient four-chambered heart, and a muscular sheet (the diaphragm) that helps to work the lungs. Most give birth instead of laying eggs, and all feed babies milk from special glands. Such soft parts do not survive as fossils, and experts distinguish fossil mammals from reptiles by differences between their bones – especially the jaws.

People disagree about how to group all mammals. Arguably the class Mammalia holds four main groups or subclasses, two extinct. Eotheria ("dawn mammals") were small, primitive, early mammals, known from rather scanty fossil finds. Prototheria

Mammal family tree
This family tree shows all 39 orders of the class Mammalia (mammals). Relationships of some are largely guesswork.
A Therapsid ancestors
B Subclass Eotheria
1 Docodonta (docodonts)
2 Triconodonta (triconodonts)
C Subclass Prototheria
3 Monotremata (monotremes)
D Subclass Allotheria
4 Multituberculata
E Subclass Theria, infraclass Pantotheria
5 Eupantotheria (eupantotheres)
6 Symmetrodonta (symmetrodonts)
F Subclass Theria, infraclass Metatheria, superorder Marsupialia
7 Marsupicarnivora
8 Paucituberculata
9 Peramelina (bandicoots)
10 Diprotodonta (koalas etc)
G Subclass Theria, infraclass Eutheria (placental mammals)
11 Taeniodontia (taeniodonts)
12 Edentata (edentates)
13 Pholidota (pangolins)
14 Lagomorpha (rabbits etc)

15 Rodentia (rodents)
16 Tillodontia (tillodonts)
17 Primates
18 Chiroptera (bats)
19 Dermoptera (gliders)
20 Insectivora (insectivores)
21 Creodonta (creodonts)
22 Carnivora (carnivores)
23 Cetacea (whales)
24 Amblypoda (amblypods)
25 Condylarthra (condylarths)
26 Tubulidentata (aardvarks)
27 Perissodactyla (horses etc)
28 Litopterna (litopterns)
29 Astrapotheria (astrapotheres)
30 Notoungulata (notoungulates)
31 Trigonostylopoidea
32 Xenungulata (xenungulates)
33 Pyrotheria (pyrotheres)
34 Artiodactyla (cows etc)
35 Hyracoidea (hyraxes)
36 Embrithopoda (embrithopods)
37 Proboscidea (elephants etc)
38 Sirenia (sea cows)
39 Desmostylia (desmostylans)

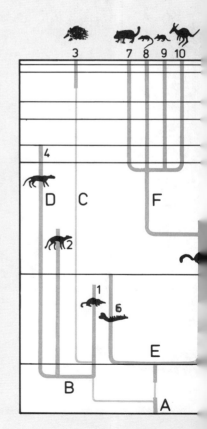

are primitive egg-layers such as the living platypus. Allotheria were early, mostly small mammals, with teeth much like a rodent's. Theria include marsupials (pouched mammals) and placentals (mammals whose babies develop in the mother well nourished by food from a special structure, the placenta).

Only small, unobtrusive mammals co-existed with the dinosaurs. When these died out, therians more than took their place, evolving shapes and sizes that suited them for life in almost every habitat. Small, early forms gave rise to bulky herbivores and carnivores. Others took to air or water. Mammals have ruled most lands for the last 65 million years, the Cenozoic Era.

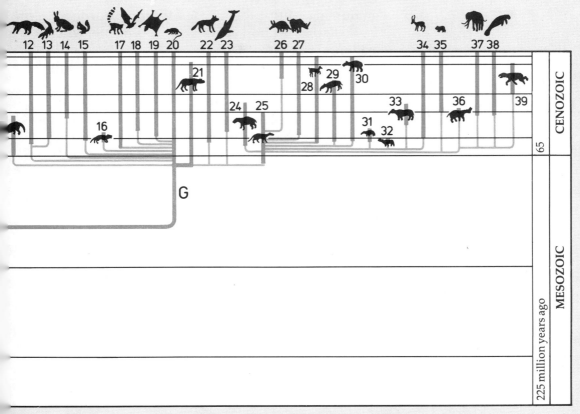

Early mammals

Of most known early mammals little more than teeth or jaws survive. Eotheria, the earliest of all, were mostly small and shrew-like Triassic and/or Jurassic creatures. Some had traces of the old, reptilian type of jaw joint. Their two main subdivisions, cocodonts and triconodonts, are named from differences in their types of teeth.

A 175-million-year gap separates the Eotheria from their possible descendants the Prototheria. These egg-laying mammals are known only from the monotremes, an order that includes the spiny anteater.

The Allotheria formed a long line of early mammals with multicuspid molars like rodents' (cusps are points on a tooth's grinding surface). Seemingly these were the first plant-eating mammals. Multituberculates, their only order, persisted 90 million years, into the Eocene, when rodents took their place.

The Theria ("true" mammals) include not only modern mammals but their fossil ancestors the pantotheres, with two orders: eupantotheres and symmetrodonts. Eupantotheres had complex molars shaped to shear and crush. Symmetrodonts had simpler teeth. Both died out by Mid Cretaceous times (some 100 million years ago).

Mammal features
Unlike most reptiles, fossil mammals show these and other features.
a Distinctive jaw hinge
b Only one bone on each side of the lower jaw
c Three auditory bones in each ear, inside the skull
d Teeth only on jaw rims
e Complex cheek teeth, with two roots or more
f A single bony nasal opening
g Growing bones show epiphyses – ends separated by cartilage from the main parts of the bones
h Enlarged braincase
i No "third eye" in the skull
j No ribs in neck or lower back area
k Limbs hinged to move fore and aft below the body
l Distinctive shoulder girdle
m Distinctive hip bones

1 **Megazostrodon,** one of the first mammals, was a tiny shrew-like triconodont with a slim lower jaw. Length: 10cm (4in). Time: Late Triassic or Early Jurassic. Place: South Africa.

2 **Echidna,** a living spiny anteater, here stands for the monotremes of which the only early fossils known are 15-million-year-old teeth from Australia. Some bones are primitive and reptilian. Like reptiles, monotremes expel all body wastes and give birth through one hole. They lack ears and have poor temperature control.

3 **Taeniolabis,** a beaver-sized multituberculate, had a heavy skull, strong jaw muscles, and big, chisel-like teeth. It probably ate nuts. Time: Early Palaeocene. Place: North America.

4 Unnamed eupantothere. This agile, squirrel-like climber lived in Late Jurassic Portugal. Its skeleton was the first found (in 1977) for any mammal dating from that time.

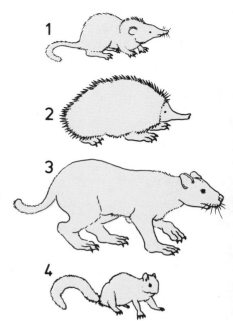

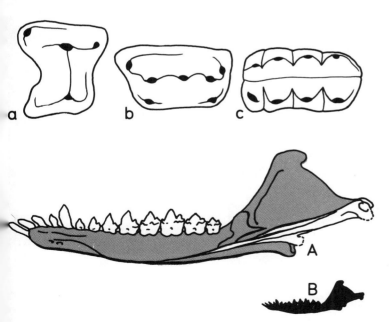

Teeth of early mammals
Almost microscopically tiny teeth are the only known remains of many early fossil mammals. Shown here are crown views of three distinctive types of upper cheek teeth (mostly much enlarged).
a Docodont tooth shaped like a dumb-bell
b Triconodont tooth, with three cusps in a row
c Multituberculate tooth, with many cusps

Distinctive jaw bones
Shown here much enlarged (**A**) and actual size (**B**) is the jaw of the early mammal *Morganucodon*. Like reptile jaws this still comprises several bones, but the dentary (shaded) is by far the largest. In later mammals only this bone persists.

153

Pouched mammals

Marsupials (pouched mammals) give birth to tiny undeveloped young. Many grow up in a pouch located on the mother's belly and supported by so-called marsupial bones. Such bones and the teeth help experts to identify fossil marsupials.

Marsupials probably evolved from eupantotheres in Mid Cretaceous times (100 million years ago). They reached all continents except Africa and Asia, but only in the island continent Australia did they escape competition from advanced placental mammals. Placental counterparts abound among the four marsupial orders. Marsupicarnivores, perhaps the parent stock, include extinct cat-like and wolf-like flesh-eaters. Paucituberculates are the insect-eating opossum rats and marsupial moles and mice. Peramelines comprise bandicoots – Australia's long-snouted rabbits. Diprotodonts, the main surviving group, include herbivorous kangaroos, koalas, wombats, and their extinct relatives – some of "giant" size, notably numbers 2–4 of the four extinct marsupials shown here (right).

Marsupial features
The American opossum shows primitive marsupial features not found in placental mammals.
a Relatively small braincase
b Holes in the bony palate (the roof of the mouth)
c Four molar teeth on each side of each jaw
d Shelf on the outer rim of the cheek teeth
e No full set of milk teeth
f Inturned lower jaw below the jaw hinge
g Marsupial (hip) bones
h Clawed second and third toes on the hind foot form a comb for grooming fur

1 **Thylacosmilus** was a leopard-sized South American marsupicarnivore equivalent of the placental sabre-toothed cat: with long, curved, stabbing upper canines sheathed in the lower jaw. One of the last in its family (the borhyaenids), *Thylacosmilus* dates from Pliocene times.

2 **Thylacoleo,** an almost lion-sized marsupial "lion" from Pliocene and Pleistocene Australia, had unique tusk-like incisors and huge premolar cheek teeth that sheared like scissors. It might have eaten flesh or fibrous fruits and tubers.

3 **Sthenurus** was a giant kangaroo with short jaws, short tail, and huge fourth toe. It stood about 3m (10ft) high and browsed on trees. Time: Pliocene and Pleistocene. Place: Australia.

4 **Diprotodon,** the largest known marsupial, was hippo-like, with big incisor tusks, but cheek teeth like kangaroos'. Length: 3.4m (11ft). Time: Pleistocene. Place: Australia.

Australia's "hippo"
Diprotodon munched plants on salt flats at Lake Callabonna in south-east Australia. Fossil finds show that individuals fell through a dried salt crust and drowned in mud beneath.

Placental pioneers

Placental mammals (the infraclass Eutheria) are so named from the pregnant females' placenta – the organ nourishing young in the womb. Placentals evolved from eupantotheres in Mid Cretaceous times. Advanced brains, and babies already well developed at birth gave placentals major advantages. In the last 65 million years they have produced 95 per cent of all known mammal genera, past and present. Here we look briefly at five early orders of placentals, two extinct.

The firstcomers were insectivores – small, nocturnal insect-eaters, some ancestral to those living insectivores shrews and hedgehogs. Closely related orders include bats (chiropterans); gliding mammals (dermopterans); the large, rat-like taeniodonts; and the bear-like tillodonts. The last two died out in Eocene times. We give one prehistoric example from each group.

Three different designs

Here we show basic differences in body structure between three groups of living mammals: monotremes (**A**), marsupials (**B**), and placentals (**C**). All release body waste from gut and bladder, and all produce egg cells in ovaries. But wastes and eggs or babies exit differently.

a Ovaries   **e** Anal canal
b Fallopian tubes   **f** Cloaca
c Uterus(es)   **g** Vagina
d Bladder   **h** Urethra

One exit (A)
In monotremes, such as the platypus, uteruses, bladder, and gut lead to a cloaca – a common exit by which eggs, liquid waste, and solid waste all leave the body. Hence the name monotreme, meaning "one opening".

Two exits (B)
In marsupials, such as the wallaby, uteruses and bladder lead to a cloaca but the gut's anal canal ends separately. Babies and liquid waste leave the body through the cloaca. Solid waste leaves from the gut's anal canal.

Three exits (C)
In placental mammals, such as the shrew, uterus, bladder, and gut each have a separate body exit. Babies leave via a vagina, liquid waste via a urethra, solid waste via the anal canal.

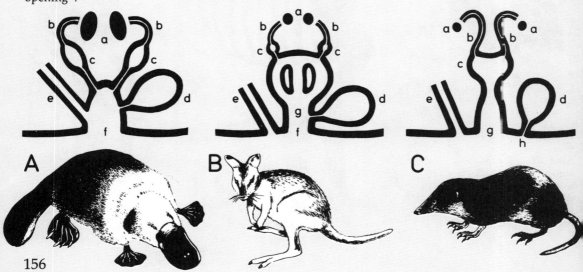

1 **Zalambdalestes** was an agile insectivore, with a long face, large eyes, small brain, long, sharp incisor teeth, "old-fashioned" molars, clawed feet, and a long tail. It belonged to the proteutherians: a suborder close to the ancestry of all main placental groups. Length: 20cm (8in). Time: Late Cretaceous. Place: Mongolia.

2 **Planetetherium,** a squirrel-sized dermopteran, glided between trees on skin webs joining legs and tail. Time: Palaeocene. Place: Montana, USA.

3 **Icaronycteris,** the first known bat, lived 50 million years ago in (Eocene) Wyoming, USA. Order: Chiroptera (bats).

4 **Stylinodon,** a taeniodont leaf-eater, resembled a huge rat with short, strong limbs, short toes, and powerful claws. Teeth were high-crowned, rootless pegs. Time: Mid-Eocene. Place: North America.

5 **Trogosus,** a tillodont, resembled a big bear, with flat, clawed feet but chisel-shaped incisors like a rodent's, and low-crowned molars. It was a herbivore. Time: Mid Eocene. Place: North America.

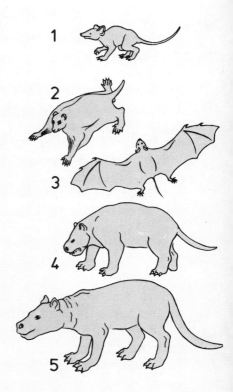

©DIAGRAM

The first fossil bat
Shown is a fossil of the first known bat, *Icaronycteris*. Fingers held up skin-membrane wings as in modern bats, and like them it slept upside down. Even such specialized placentals shared the following common features.
a Relatively big brain case
b Solid bony palate
c Lower jaws lacking inturned angle
d No marsupial hip bones
e Distinctive teeth

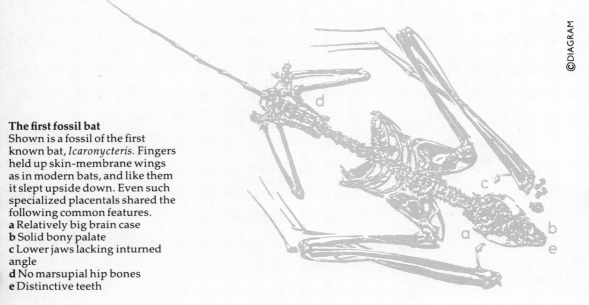

Primates 1

Primates – man, apes, monkeys, and their kin – include the most intelligent of all animals. Yet their direct ancestors were insectivores, the earliest, most primitive placental mammals. Primates primitively kept five toes and fingers, but gained grasping hands, forward-facing eyes, and a big "thinking" region of the brain. Eyes and hands co-ordinated by the brain adapted early primates for an agile life among the trees. Meanwhile changes in the teeth adapted them for eating fruits or most other foods.

Primates resembling today's tree shrews probably appeared by 65 million years ago, giving rise to several groups. First came the relatively small and small-brained prosimians, including ancestors of today's lemurs and tarsiers. These were plentiful 60 to 40 million years ago in North America, Eurasia, and Africa. Prosimians led to the brainer and often bigger anthropoids (monkeys, apes, and men) that largely took their place. Here are examples from the primates' five suborders.

1 **Plesiadapis,** a squirrel-sized, squirrel-like prosimian, had a long snout, side-facing eyes, big chisel-like teeth, long bushy tail, and claws not nails. It could not grasp with its hands. It ate leaves, largely on the ground. Time: Mid Palaeocene. Place: North America and Europe. Suborder: Plesiadapoidea (extinct by Mid Eocene times).

2 **Adapis** had a shorter snout than *Plesiadapis*, forward-facing eyes, relatively bigger brain, and grasping hands and feet with nails not claws. It climbed forest trees, eating shoots, fruits, eggs, and insects. Length: 40cm (16in). Time: Mid–Late Eocene. Place: Europe. Suborder: Lemuroidea (lemurs), probably ancestral to apes.

3 **Necrolemur** was small with huge eyes, big ears, a small "pinched" nose, long tail, and long tree-climber's limbs with gripping toe pads. It was an Eocene European ancestor of modern tarsiers. Suborder: Tarsioidea.

4 **Dolichocebus** was small and squirrel-like with a bushy tail, and claws not nails. It lived in Oligocene South America. Suborder: Platyrrhini ("flat noses") – New World monkeys. These have widely spaced, outward facing nostrils and long tails, and lack good thumb-and-finger grip.

5 **Mesopithecus** was probably ancestral to the slim, long-legged, long-tailed monkeys known as langurs. It lived in Late Miocene Greece and Asia Minor. Suborder: Catarrhini ("down-facing noses") – Old World monkeys, apes, and men. These have nostrils close together, good thumb-and-finger grip, and relatively big brains; tails become short in many later monkeys and are absent in apes and man.

Primate features
These clues help experts recognize a primate fossil.
a Large brain case
b Jaws often short
c Low-crowned cheek teeth with low, blunt cusps (designed to cope with almost any food)
d Usually no third molar or first premolar tooth
e Orbits (eye sockets) large and facing forward
f Bony bar behind the orbits
g Small nasal opening
h Long, flexible limbs
i Five toes and fingers, usually with nails not claws
j Gap between thumb and big toe and other digits – an aid in grasping

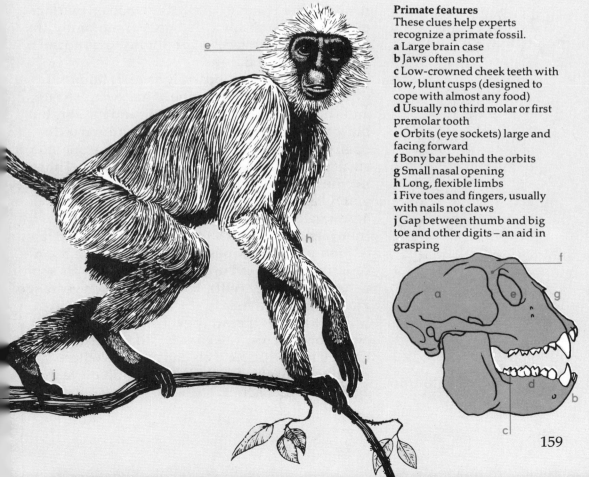

159

Primates 2

Hominoids, the catarrhine superfamily containing apes and men, arose from prosimian ancestors some 35 million years ago, probably in Africa. Bigger and brainier than monkeys, apes spread through Europe and Asia 25–10 million years ago.

Four hominoid families evolved: oreopithecids, a dead end; pongids (great apes); hylobatids (gibbons); and hominids (men and ape men). Our examples stress those that led to modern man.

1 **Aegyptopithecus,** from Oligocene Egypt (26–28 million years ago) was a small ape with a short tail. It had low brows and jutting snout but bony eye sockets not just bars like early primates. It might have led to modern apes.

2 **Dryopithecus** was a chimpanzee-like ape with relatively short arms. It probably stood on two legs but climbed on all fours. Its Miocene subfamily, the dryopithecines, lived 25–10 million years ago in Africa, Asia, and Europe.

3 **Ramapithecus** was possibly an early hominid evolved from a dryopithecine. It had a flattish face, and "human" teeth: small canines and incisors but big, crushing, grinding molars. It ventured on to open grassland, maybe using sticks and stones to kill small prey, or in defence. Height: 1.2m (4ft). Time: maybe 15–7 million years ago. Place: Africa, Europe, and Asia.

4 **Australopithecus,** an ape man, possibly evolved from *Ramapithecus.* It had an ape-like skull and face, human-type teeth, one-third modern man's brain capacity, but walked upright and probably ate some meat. Height: 1.2m (4ft). Time: 5–1 million years ago. Place: Africa and Asia.

5 **Homo habilis,** possibly evolved from a slender australopithecine, looked rather like a small-brained child. It made pebble tools and maybe shelters. Height: 1.2m (4ft). Time: about 4–2 million years ago. Place: East Africa.

6 **Homo erectus** had thick eyebrow ridges, heavy teeth, and receding chin and forehead, but a bigger brain and larger body than *Homo habilis*, its likely ancestor. It pioneered the use of fire, standard (hand axe) tools, and big-game hunting. Height: 1.7m (5ft 6in). Time: 1,750,000–200,000 years ago. Place: Africa, Asia, and Europe.

7 **Homo sapiens,** our own species, evolved about 200,000 years ago, probably from *Homo erectus*. Key innovations included larger brain, higher, more rounded forehead, and increased body size. We show the Neanderthal subspecies of 100,000–35,000 years ago. Our own subspecies (*Homo sapiens sapiens*) appeared perhaps only 40,000 years ago.

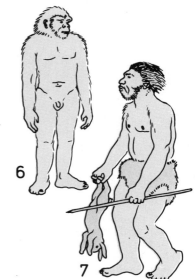

©DIAGRAM

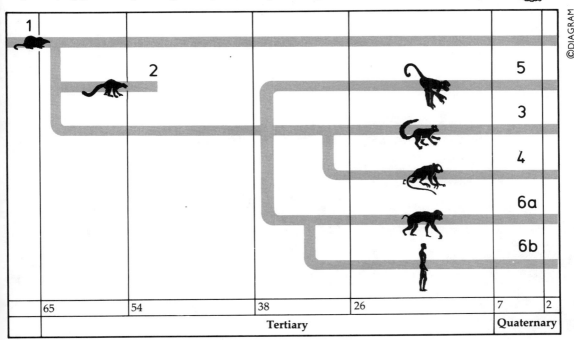

65	54	38	26	7		2
		Tertiary				Quaternary

Primates' family tree
This shows: insectivores ancestral to the primates (**1**); the primates' five suborders (**2–6**); and the catarrhines' superfamilies including apes and men (**a–b**).

1 Insectivores
2 Plesiadapoidea (early primates)
3 Lemuroidea (lemurs and lorises)
4 Tarsioidea (tarsiers)
5 Platyrrhini (New World monkeys
6 Catarrhini
a Cercopithecoidea (Old World monkeys)
b Hominoidea (apes and men)

161

Creodonts

Creodonts were the first successful flesh-eating placental mammals. Many walked "flat footed" on short, heavy limbs tipped with claws. The tail was long, the brain was small, and the teeth were less efficient than a cat's for stabbing flesh or shearing through it. Creodonts ranged from weasel size to beasts bigger than a bear. Small species might have eaten insects. Larger kinds probably included big-game hunters, carrion eaters, and omnivores. Creodonts roamed northern continents about 65–5 million years ago – as "top dogs" in the Eocene (54–38 million years ago). Then there were still hoofed animals slow enough for them to catch, and they lacked competition from brainier, more lethal killers than themselves.

There were two creodont suborders: the early deltatheridians and the hyaenodonts. The hyaenodonts were the main group, with two families: oxyaenids and hyaenodontids.

A flesh-eating monster
Megistotherium might have been the largest-ever flesh-eating mammal. This huge creodont weighed about 900kg (1980lb). The head was twice as large as any bear's and was armed with mighty canine teeth. *Megistotherium* quite likely killed elephant-like mastodons. Miocene rocks in Libya contain this monster's fossils.

1 **Deltatheridium,** a deltatheridian, had creodont-like teeth (but just might have been a marsupial). Length: maybe 15cm (6in). Time: Late Cretaceous. Place: Mongolia.

2 **Patriofelis** was a bear-size oxyaenid. Oxyaenids had a short, broad head, deep strong jaws, and cheek teeth that crushed rather than sheared. Time: Mid Eocene. Place: North America.

3 **Hyaenodon** included wolf-sized species that tackled big, hoofed animals. It belonged to the hyaenodontids – a far bigger, longer-lasting group than oxyaenids, and usually with a longer head, more slender jaws, a slimmer body, longer legs, and a tendency to walk on the toes. Teeth sheared rather than crushed. *Hyaenodon* lived in northern continents from Late Eocene to Miocene times.

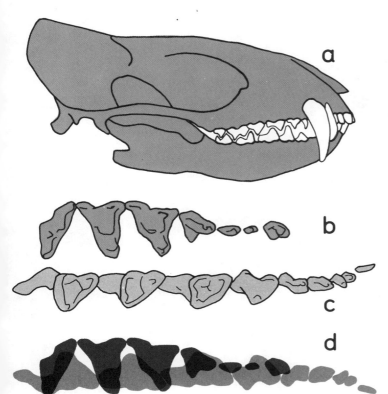

Teeth to shear and chop
Early mammals had cheek teeth with triangular crowns. Those in early creodonts worked to shear and chop, as shown here for *Deltatheridium*.
a Skull with teeth, shown enlarged.
b Crown view of upper cheek teeth, shown much enlarged.
c Crown view of lower cheek teeth, shown much enlarged.
d Upper and lower cheek teeth here shown meshed together. The front edges of the lower teeth slid along the back edges of the upper teeth ahead of them. The low "heel" behind the main part of each lower tooth fitted into the surface of the upper tooth behind it, stopping further sliding. Partial meeting of cusps ("peaks") on the tooth crowns helped chop up food.

Modern carnivores 1

Modern carnivores form one order, Carnivora, with two suborders: fissipeds or "split feet" (dogs, cats, and their kin) and pinnipeds (the "fin footed" eared seals, earless seals, and walruses).

Fissipeds had bigger brains, keener ears, more deadly piercing canines and shearing cheek teeth, and longer limbs than creodonts. Cunning, agile fissipeds replaced creodonts as the major carnivores, for only fissipeds could catch and kill new kinds of speedy herbivores that replaced slower types.

Fissipeds evolved from insectivores by 60 million years ago. By 35 million years ago they dominated life on most continents. Three superfamilies arose: first miacids, then their descendants aeluroids or feloids (civets, hyaenas, and cats) and arctoids or canoids (dogs, bears, and their kin). Beasts from the first two groups are pictured on this spread.

Victor and victim
Big stabbing cats like *Smilodon* probably preyed on big, slow-moving herbivores like the large, prehistoric ground sloth *Mylodon*. Remains of both occur in Los Angeles' famous La Brea tar pits. Tar trapped big herbivores that drank rainwater concealing tar beneath. The creatures' struggles attracted predators.

1 **Miacis** was a small, weasel-like, tree-climbing miacid. Length: 60cm (2ft). Time: Eocene. Place: North America, Europe, Asia.

2 **Genetta,** the genet, resembles Eocene members of its Old World civet family (viverridae). It is a cat-like forest dweller with short limbs, long tail, and retractile claws. Length: 96cm (3ft 2in). Time: Pleistocene on. Place: Africa, Eurasia.

3 **Percrocuta** had a big head, strong jaws, bone-cracking teeth, and longish legs. Species of this early hyaena were wolf size to lion size. Time: Miocene. Place: Africa and Asia.

4 **Dinictis** was a puma-sized ancestor of biting and stabbing cats (all in the felid family). Time: Early Oligocene–Early Miocene. Place: North America.

5 **Felis leo spelaea,** the great "cave lion", was a biting cat one-third larger than the largest lion. It lived in Mid Pleistocene Europe.

6 **Smilodon,** the "sabre-tooth tiger", was a heavy, lion-sized stabbing cat. Its dagger-like upper canines stabbed large prey or slashed blood vessels in their necks. Time: Late Pliocene–Pleistocene. Place: North and South America.

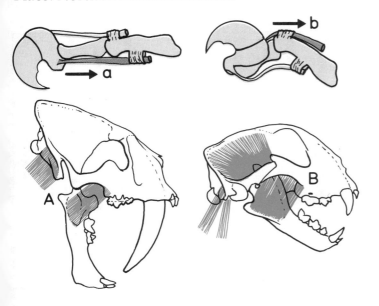

Cats' claws
Ability to unsheathe claws and spread toes provides most cats with formidable weapons.
a Unsheathing: pulling a toe's lower tendon makes a claw curve down and out.
b Sheathing: pulling a toe's upper tendon makes a claw curve up and back.

Stabbing and biting
Stabbing cats and biting cats evolved different techniques for using teeth as weapons.
A Stabbing cats like *Smilodon* used strong neck muscles to strike downward with the head.
B Biting cats like the cheetah have strong jaw muscles and exert a formidable bite.

©DIAGRAM

165

Modern carnivores 2

Arctoids (or canoids) evolved from miacids over 40 million years ago. These carnivores mostly share a skull peculiarity involving the middle ear. They have less specialized teeth than cats and cannot sheathe their claws. Some eat almost any food. Four families evolved. Canids include dogs, wolves, and extinct bear-dogs. The primitive mustelids include weasels, otters, and badgers. Ursids are bears and the giant panda. Procyonids are raccoons, coatis, kinkajous, and lesser pandas.

By 25 million years ago arctoids gave rise to pinnipeds – seals and walruses. This suborder of aquatic carnivores developed sleek, streamlined bodies, webbed limbs, and teeth designed for catching fish or (in walruses) for crushing clams. But unlike whales or ichthyosaurs, seals kept a flexible neck and the tail is reduced to a stump.

1 **Cynodictis,** a fox-sized early dog, had a long, low body, long neck, long tail, sharp shearing cheek teeth, and longer legs and larger brain than miacids. Time: Late Eocene–Early Oligocene. Place: Europe and East Asia.

Carnivore family tree
This shows likely links between families of the order Carnivora.
1 Miacid fissipeds
a Miacidae
2 Aeluroid fissipeds
b Ursidae (bears)
c Canidae (dogs)
d Procyonidae (raccoons)
e Mustelidae (weasels)
3 Pinnipeds
f Odobenidae (walruses)
g Otariidae (eared seals)
h Phocidae (earless seals)
4 Aeluroid fissipeds
i Hyaenidae (hyaenas)
j Viverridae (civets)
k Felidae (cats)

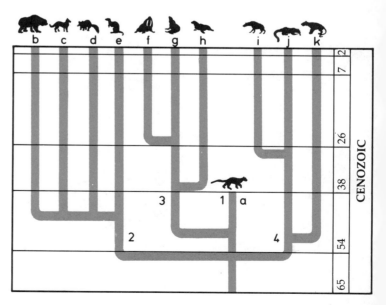

166

2 **Daphoenodon,** the largest canid of its day, was a wolf-like "bear-dog" from Early Miocene North America. It had rather short limbs, spreading toes, a long, strong tail, and was perhaps omnivorous.

3 **Osteoborus,** a hyaena-like canid scavenger, had a bulging forehead and strong jaws. It lived in Miocene North America.

4 **Megalictis** from Early Miocene North America resembled a wolverine but was as big as a black bear. It was the largest-ever mustelid.

5 **Ursus spelaeus** the great "cave bear" of Pleistocene Europe, was descended from the dog family. Bears walk flat footed, not on toes like dogs, and most are omnivores. Length: 1.6m (5ft 3in).

6 **Phlaocyon** was a small, raccoon-like procyonid (some say canid) from Miocene North America. It climbed trees, had grinding but not shearing cheek teeth, and was omnivorous.

7 **Allodesmus,** an early relative of sea-lions, probably resembled a sea-elephant, the largest earless seal alive today. Time: Early Miocene–Mid Miocene. Place: North Pacific Ocean.

Formidable jaws
Here we compare skulls of a cave bear (**A**) and a fox (**B**), drawn to scale. Cave bears were omnivores, despite their massive jaws.

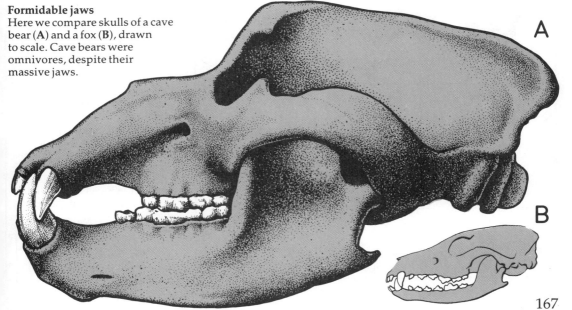

Condylarths

These were the first ungulates (hoofed mammals), abundant 65 to 40 million years ago. Early kinds had claws, not hooves or nails, and some ate flesh, not plants. Later kinds developed chopping, grinding teeth for pulping leaves, and long limbs tipped with nails or hooves, for running fast away from carnivores. The pioneers were rabbit-sized; some later forms were longer than a large bear.

Condylarths evolved from small insectivores, and probably gave rise to all the more advanced hoofed mammals (maybe also to the strange, long-snouted aardvark). Condylarths spread through northern continents and into South America and Africa. Most fossils come from Palaeocene rocks in North America, and from Eocene rocks there and in South America, Europe, and Asia. We give examples from contrasting families in the order Condylarthra.

Early ungulates
This family tree shows likely relationships between both orders of primitive hoofed mammals described on pages 168–171. The modern aardvark may be descended from the condylarths.
1 Condylarthra (condylarths)
2 Amblypoda (amblypods)
a Pantodonta (pantodonts)
b Dinocerata (uintatheres)
3 Tubulidentata (aardvarks)

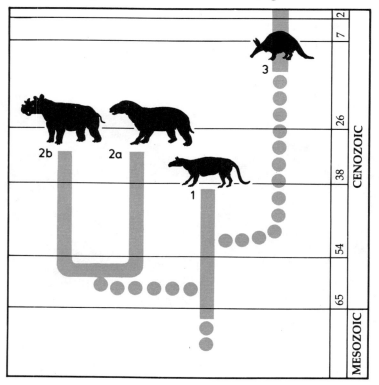

Early ungulate features (right)
The skull and feet bones of *Phenacodus* show some features found in early carnivores, but others stressed in herbivores like sheep and horses.
a Long, low skull (shown half actual length), resembling that of early carnivores
b Large canine teeth, as in early carnivores
c Ridged, square-crowned molars forming a battery of cheek teeth capable of crushing vegetable food
d Bones of hind foot, tipped with small, blunt hooves not nails
e Bones of forefoot also tipped with hooves

© DIAGRAM

1 **Protungulatum,** the first-known ungulate, was
rabbit-sized. It ate plants and maybe other foods.
Time: Late Cretaceous–Early Palaeocene. Place:
North America. Its arctyonid family featured early,
mostly small condylarths with a supple back, short
limbs tipped with claws, a long tail, long, low head,
and "old-fashioned" molars with triangular crowns.

2 **Andrewsarchus** was a huge, heavy, bear-like
condylarth, probably omnivorous. Length: 4m (13ft).
Time: Late Eocene. Place: East Asia. Family:
Mesonychidae, largely dog-like condylarths. Many
had blunt-cusped cheek teeth capable of crushing
bones, and ran dog-like on toes with flattened nails
not claws. Packs maybe hunted big herbivores.

3 **Phenacodus,** a phenacodontid condylarth, was a
sheep-sized plant-eater, with big canine teeth and
square-crowned molars. It was the earliest known
mammal with hooves not nails or claws. Fossil
skeletons show that it roamed woods and
shrublands in North America and Europe. Time:
Late Palaeocene–Early Eocene.

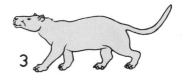

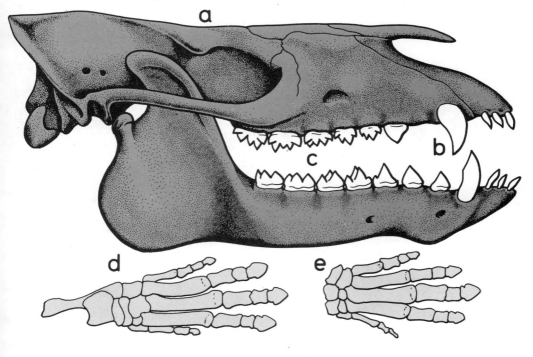

Amblypods

Amblypods ("slow footed") were early, ponderous hoofed herbivores with broad, low, ridged cheek teeth. Some had tusks or horns. Many might have lived in swamps. Size, weapons, or habitat helped save them from attack by creodonts.

Amblypods flourished 60 to 30 million years ago, mostly in Palaeocene and Eocene North America, but in Europe and East Asia too.

There were two suborders. Pantodonts ranged from dog size to pony size. They had short legs and short, broad feet. Canine teeth were long and cheek teeth had a simple pattern.

Uintatheres included rhino-sized mammals – among the largest of their age. Many had strange bony horns and long, wicked-looking upper canines. Upper molars bore a broad, v-shaped crest.

Here we show three pantodonts from different families, and the best known uintathere.

The Uinta beast
Uintatherium's low-crowned cheek teeth, massive post-like limbs, and stubby toes suggest it browsed on soft-leaved plants in grasslands and along the woodland edge.
Fossil hunters found its bones where Colorado's Uinta Mountains jut north toward Wyoming.

1 **Pantolambda,** was one of the first hoofed mammals as big as a sheep. It had heavy legs, short feet, and a long, low head with large canine teeth. It might have wallowed, hippo-like, and browsed on land. Time: Mid Palaeocene. Place: North America and Asia.

2 **Barylambda** was pony-sized, with heavy body, small head, and primitive teeth. It might have sat on its haunches to browse high up. It lived in Late Palaeocene–Early Eocene North America.

3 **Coryphodon** had a long, heavy body, large head, wide muzzle, and knife-like upper canines. Length: 3m (9ft 10in). Some think it lived largely in water. Time: Late Palaeocene–Early Eocene. Place: northern continents.

4 **Uintatherium** was a massive uintathere the size of a large rhinoceros. Thick limbs with spreading feet held up its heavy body. Three pairs of bony knobs sprouted from the head, and males had wickedly long, strong, upper canines. Time: Eocene. Place: North America.

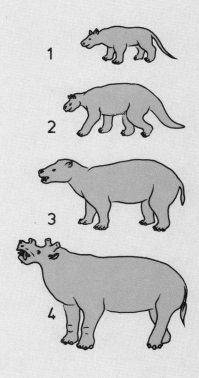

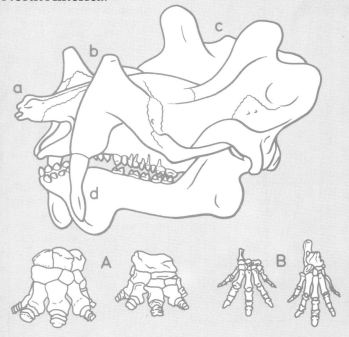

A "six-horned" head
Uintatherium's great grotesque skull 76cm (30in) long shows weapons used by jousting males or in defence.
a Two small knobs on the nose
b Two larger knobs between the nose and eyes
c Two big broad knobs above and behind the eyes
d Huge upper canines, protected by a flange in the lower jaw

Two kinds of feet
Big heavy herbivores tended to develop much more massive foot and toe bones than their smaller, lighter relatives.
A The short, thick, strong bones that bore the weight of mighty *Uintatherium*
B The relatively slim, thin bones of *Pantolambda's* fore and hind feet

Subungulates

These "not quite ungulates" comprise seemingly quite unrelated hyraxes, sirenians (sea cows), proboscideans (elephants and their kin), and the extinct embrithopods and desmostylans.

Close study of their bones and teeth in fact reveal relationships. Subungulates tend to lack a collar bone but keep five-toed feet, with nails rather than hooves. Many have few front teeth, though a pair often become tusk-like, also enlarged premolars and crosswise ridges on the grinding cheek teeth. Their yet unknown ancestors might have lived in swamps in Africa some 60 million years ago. We show examples from four of the five orders of subungulates.

1 **Saghatherium** was an agile, rodent-like hyrax with chisel-like incisors but toes with hoof-like nails. Time: Early Oligocene. Place: Egypt. Length: 40cm (16in). Living hyraxes are also small but one prehistoric kind was rhinoceros size.

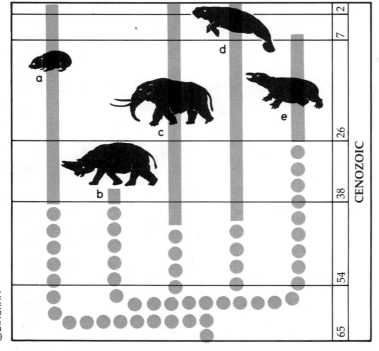

CENOZOIC

Subungulate family tree (left)
This family tree shows likely relationships between the five subungulate orders, of which two (**b** and **e**) are extinct.
a Hyracoidea (hyraxes)
b Embrithopoda (embrithopods)
c Proboscidea (elephants and their ancestors)
d Sirenia (sea cows)
e Desmostylia (desmostylans)

Mystery monster (right)
Twin-horned *Arsinoitherium* had a big head, and limbs like elephants'. The cross-crested, high-crowned cheek teeth were better than those of its contemporaries for crushing tough-leaved plants. Lack of known fossil relatives makes this creature's origin something of a mystery.

2 **Arsinoitherium,** from Early Oligocene Egypt, is the only known embrithopod. This rhinoceros-sized mammal had a huge pair of horns side by side on its head. It probably munched coarse-leaved plants in swamps. Length: 3.4m (11ft).

3 **Desmostylus,** a heavy, walrus-like desmostylan, swam or waded on North Pacific coasts from Early Miocene to maybe Early Pliocene times. Short tusks grew forward from both jaws. Rear teeth were close-packed, heavily enamelled bundles, perhaps used for crunching shellfish or munching seaweed.

4 **Protosiren** was an early sea cow about 2.4m (8ft) long, from Middle Eocene North Africa and Europe. Sea cows developed broad, beak-like snouts, flipper-like front limbs, and a broad, flat tail; they lost hind limbs. Living species browse on aquatic plants in tropical river mouths.

Proboscideans 1

Elephants, their ancestors, and other kin make up the subungulate order Proboscidea ("long snouted"). The first pig-sized, trunkless proboscideans lived in Africa 40 million years ago. From Africa, their descendants invaded all continents except Australia and Antarctica. Meanwhile they grew in size. Many were immense slab-sided beasts with "tree-trunk" limbs, and a huge head armed with tusks and brandishing a flexible muscular trunk – a "hand" bringing leaves and water to the mouth. In four pages we show something of the astonishing variety of prehistoric proboscideans.

Our examples come from three of their four suborders: moeritheres, deinotheres, and euelephantoids (mastodonts and elephants). The barytheres from Eocene Egypt are little known.

1 **Moeritherium,** a pig-sized, heavy footed moerithere, had a short, flexible snout and forward-jutting incisor teeth. It lived in swampy lands. Time: Late Eocene–Early Oligocene. Place: northern Africa (Egypt, Mali, Senegal).

Spade blades (below)
Platybelodon's lower jaw ended in two flat, broad tusks (here shaded) like spade blades.

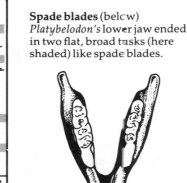

54	38	26	7	2
CENOZOIC				

Proboscidean family tree
1 Moeritherioidea (moeritheres)
2 Deinotherioidea (deinotheres)
3 Euelephantoidea (mastodonts and elephants)
a Gomphotherioidea (long-jawed mastodonts)

b Mastodontoidea (crested toothed mastodonts)
c Stegodontoidea (stegodonts)
d Elephantoidea (mammoths and elephants)
4 Barytherioidea (barytheres)

2 **Deinotherium** belonged to the deinotheres: beasts up to 4m (13ft) high, with down-curved tusks in the lower jaw – forks for digging roots perhaps. Time: Miocene–Pleistocene. Place: Africa and Asia.

3 **Phiomia,** an early mastodont ("nipple toothed" proboscidean), had pig-like cheek teeth, long lower jaw, four short tusks, and short trunk. Height: 2.4m (8ft). Time: Early Oligocene. Place: Egypt.

4 **Platybelodon,** a mastodont "shovel tusker", dug up plants with broad blade-like teeth jutting from its long lower jaw. Time: Miocene. Place: Asia and North America.

5 **Mastodon** had long, curved upper tusks, strong crests across its cheek teeth, and a coat of reddish hair. It browsed in forests of Pleistocene North America, dying off 8000 years ago.

6 **Stegodon** was a long-tusked possible ancestor of mammoths (see next page). Time: Pliocene– Pleistocene. Place: Asia and Africa.

A living bulldozer
Platybelodon might have used its tusks to scoop up land- or water-plants as shown here.

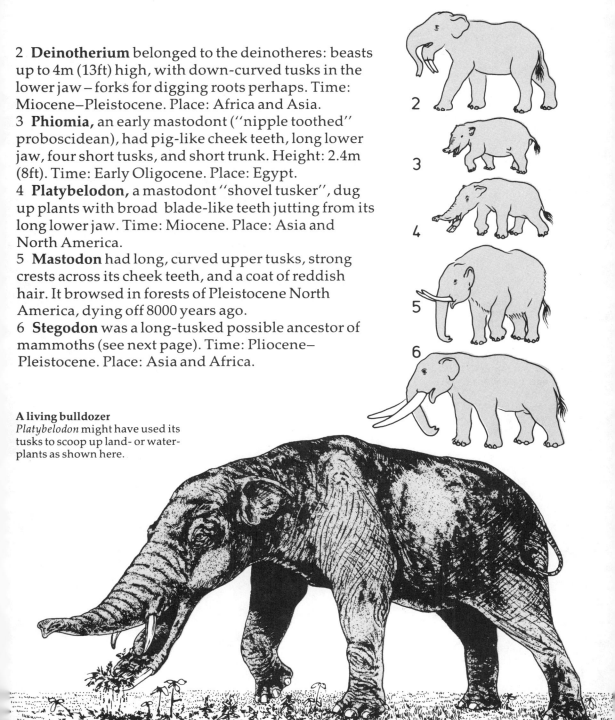

175

Proboscideans 2

Mammoths ("giants") included all extinct members of the Elephantinae, the elephant subfamily. They evolved from mastodons (see page 175) but were mostly taller, with higher skulls, shorter jaws, and more complex molar teeth, designed for grazing. Some were the largest-ever elephants; some, dwarfs. Cold-adapted types were hairy.

Mammoths appeared in Africa five million years ago, colonized the Northern Hemisphere, and died out about 10,000 years ago, after giving rise to modern elephants. There were several genera and many species. Our examples give an idea of the range of types and sizes.

1 **Elephas falconeri** from the island of Sicily was one of several dwarfed island elephants – an agile beast only one-quarter the size of its mainland ancestors.

Mammoth skull and teeth
A woolly mammoth skull appears here to scale with a human skull.
a Huge, curved tusks served in defence and maybe helped clear snow from pasture.
b As it grew, each huge molar pushed out its predecessor(s).
c A crown view shows the many crosswise ridges that helped this molar grind like a mill.

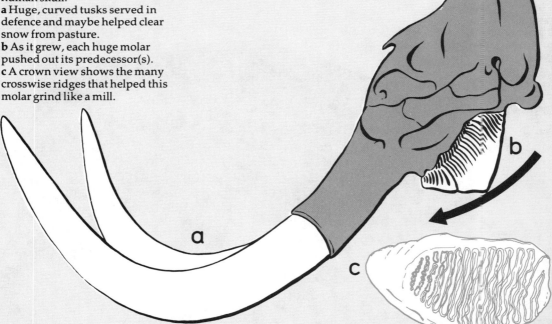

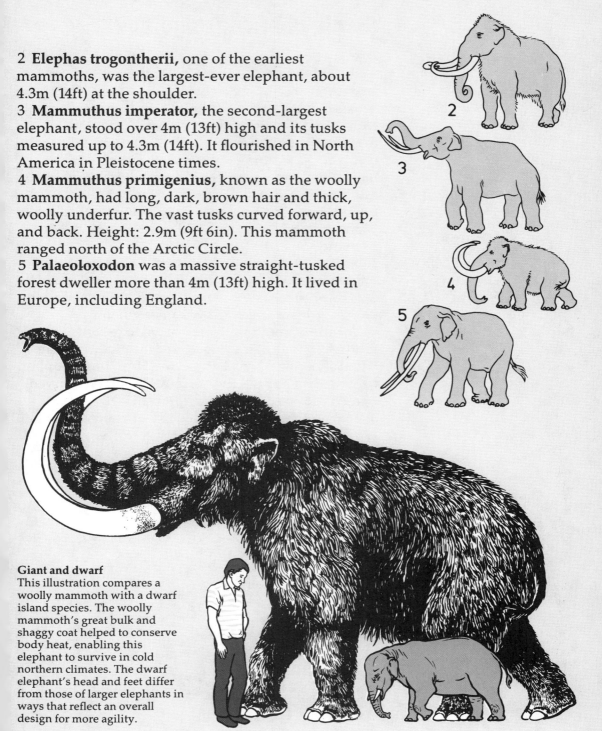

2 **Elephas trogontherii,** one of the earliest mammoths, was the largest-ever elephant, about 4.3m (14ft) at the shoulder.

3 **Mammuthus imperator,** the second-largest elephant, stood over 4m (13ft) high and its tusks measured up to 4.3m (14ft). It flourished in North America in Pleistocene times.

4 **Mammuthus primigenius,** known as the woolly mammoth, had long, dark, brown hair and thick, woolly underfur. The vast tusks curved forward, up, and back. Height: 2.9m (9ft 6in). This mammoth ranged north of the Arctic Circle.

5 **Palaeoloxodon** was a massive straight-tusked forest dweller more than 4m (13ft) high. It lived in Europe, including England.

Giant and dwarf
This illustration compares a woolly mammoth with a dwarf island species. The woolly mammoth's great bulk and shaggy coat helped to conserve body heat, enabling this elephant to survive in cold northern climates. The dwarf elephant's head and feet differ from those of larger elephants in ways that reflect an overall design for more agility.

South American ungulates

A whole "zoo" of hoofed mammals evolved in South America while sea cut it off from other lands. Many amazingly resembled rodents, horses, and other beasts that evolved elsewhere. There were six orders. Notoungulates, litopterns, astrapotheres, and trigonostylopoids probably evolved from early condylarths. Xenungulates and pyrotheres might have come from amblypods. Some endured for many million years. But none survived long once land linked North and South America about two million years ago, and carnivores and ungulates invaded from the north.

1 **Notostylops** was a small primitive hoofed mammal from Early Eocene Patagonia. It belonged to the Notoprogonia suborder of the notoungulates, the largest order of South American ungulates.

2 **Toxodon** was a rhinoceros-sized toxodont ("bow-toothed") – one of the largest and last notoungulates. It had short legs, broad, three-toed feet, big cropping incisor teeth and tall cheek teeth. The molars curved in toward each other. Time: Pliocene–Pleistocene.

3 **Protypotherium** resembled a large rodent. It was a typothere notoungulate with claws not hooves. Length: 51cm (20in). Time: Miocene.

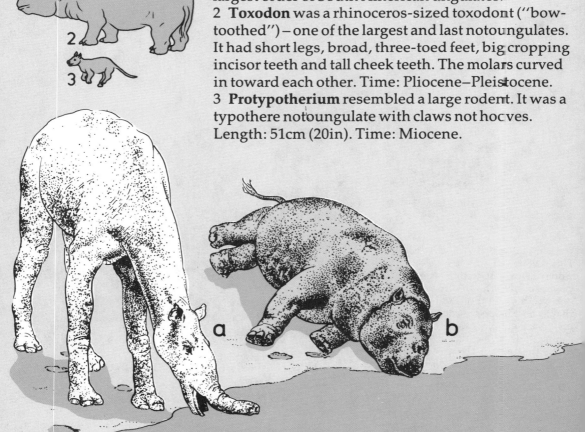

4 **Pachyrukhos** had long hind limbs and a stumpy tail. Probably it ran and even leaped like a hare. This Miocene creature belonged to the hegetotheres, the smallest notoungulates.

5 **Thoatherium** looked astonishingly like a small horse. It belonged to the proterotheriids, a family of litopterns. Time: Early Miocene.

6 **Macrauchenia** – camel-sized, and camel-like – might have had a trunk. It was a macraucheniid litoptern. Time: Pleistocene.

7 **Astrapotherium** belonged to the big astrapotheres. It had a "sawn-off" face, dagger-like canine teeth, and possibly a trunk. Hind limbs were weak. Perhaps it lived in water. Length 2.7m (9ft). Time: Oligocene–Miocene.

8 **Trigonostylops** was a small beast once grouped with astrapotheres, but now put in its own order: Trigonostylopoidea. Time: Palaeocene–Eocene.

9 **Carodnia,** the only known xenungulate, might have resembled a uintathere. Time: Late Palaeocene.

10 **Pyrotherium,** a pyrothere, amazingly resembled a large early elephant in its trunk, chisel-like tusks, and type of teeth. Time: Oligocene.

Life by a pool
This restoration shows a likely scene in South America about three million years ago. A *Macrauchenia* (**a**) drinks from a pool inhabited by two *Toxodon* (**b,c**). Alignment of the nose, eyes, and ears suggests that *Toxodon* could swim all but submerged. Perhaps this bulky mammal lived like a hippopotamus.

©DIAGRAM

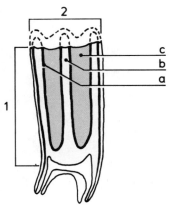

Teeth for grazing
The cheek tooth of a modern horse is designed to chew hard, abrasive grass.
1 High crown
2 Uneven grinding surface produced by wear on enamel (**a**), dentine (**b**), cement (**c**).

Evolving hooves
These illustrations show the hooves of *Hyracotherium* (**A**), *Miohippus* (**B**), *Merychippus* (**C**), and *Equus* (**D**). Notice that side toes shrink and disappear, until each foot has only an enlarged middle toe ending in a big, broad hoof. Changes in foot design – like changes in skull design – adapted horses for life as grazers on open plains.

Horses

Horses, rhinoceroses, tapirs, and others make up the perissodactyls, or "odd-toed" ungulates – one of the two great surviving orders of hoofed mammals. Evolving over 55 million years ago, perissodactyls became the most abundant ungulates, but then declined.

Few creatures left a richer fossil record than the horses – a superfamily in the hippomorph suborder. Dog-sized forest browsers living more than 50 million years ago gave rise to big, speedy grazers, with teeth for chewing tough-leaved grasses that began replacing forests. From their home in North America, horses reached Eurasia, Africa, and South America. Only members of one genus still survive.

1 Hyracotherium, the first known horse, was fox-sized. It had a short neck, curved back, long tail, slim limbs, and long four-toed forefeet and three-toed hind feet. Its low-crowned cheek teeth munched soft leaves in swampy North American and European forests. Time: Late Palaeocene–Early Eocene.

2 Mesohippus was bigger than *Hyracotherium*, with a straighter back, longer legs, three-toed forefeet, and enlarged premolar teeth. Height: 60cm (2ft) Time: Early–Mid Oligocene.

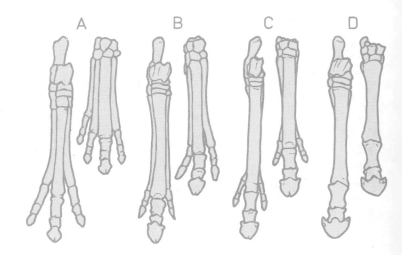

A   B   C   D

3 Miohippus was a little larger still, with big middle toe, big, molar-type premolars, and ridged cheek teeth. Time: Mid Oligocene–Early Miocene.

4 Merychippus was pony-sized, walked on each middle toe, and had a long neck and tall ridged cheek teeth for chewing grasses. Time: Miocene.

5 Pliohippus, 1.2m (4ft) tall, was the first one-toed horse. Time: Pliocene.

6 Equus, the modern horse, 1.5m (5ft) tall, evolved two million years ago, and reached most continents. Surviving wild species are Przewalski's horse, the wild ass, and zebras.

7 Palaeotherium belonged to the palaeotheriids, a family evolved from early horses. It had three hooves per foot and seemed half horse, half tapir. Height: 75cm (30in). Time: Late Eocene–Early Oligocene. Place: Europe.

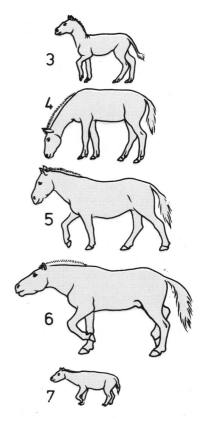

Horse trends
These two early horses show important evolutionary trends.
a *Orohippus,* a middle Eocene descendant of *Hyracotherium,* was still only whippet-sized, but its teeth could cope with tougher leaves, and it lived in drier woodlands.
b *Merychippus,* a larger horse, could eat the hard, abrasive grasses of Miocene North America. A big gap separated its cropping front teeth from grinding cheek teeth, and cement filled "pools" between the enamelled ridges of each cheek tooth's crown. When *Merychippus* ran fast, its side toes did not reach the ground.

©DIAGRAM

Brontotheres and chalicotheres

These were the strangest of all odd-toed ungulates. Brontotheres included massive beasts with elephantine limbs and a blunt, bony prong utting from the nose. They ate only soft-leaved plants. The brontothere (titanothere) superfamily was in the same suborder as horses and, like them, came from condylarths. The group flourished in all northern continents and lived about 50–30 million years ago.

Chalicotheres ranged from sheep size to cart-horse size. They looked a bit like horses, yet their three-toed feet had big, curved claws that could be sheathed. Some think they walked on knuckles, using claws to dig up edible roots or pull down leafy branches. The suborder was never plentiful but persisted more than 50 million years until less than two million years ago. Fossils come from Europe, Asia, Africa, and North America.

Perissodactyls
This family tree shows links between suborders (**b–d**) and superfamilies (**i–iv**) in the order Perissodactyla ("odd-toed" ungulates) described on pages 180–185.
a Condylarth ancestors
b Hippomorpha
i Equoidea (horses)
ii Brontotherioidea (titanotheres or brontotheres)
c Ancylopoda (chalicotheres)
d Ceratomorpha
iii Tapiroidea (tapirs)
iv Rhinocerotoidea (rhinoceroses)

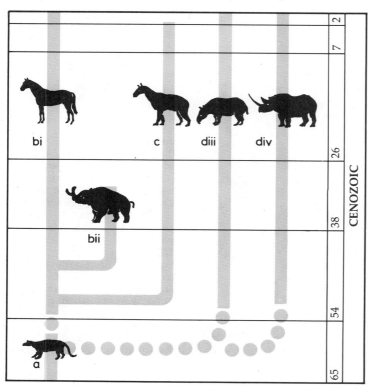

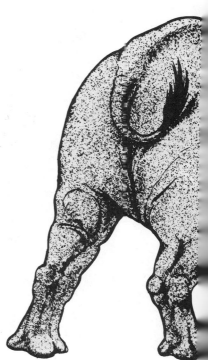

©DIAGRAM

182

1 **Brontotherium** was between a rhinoceros and elephant in size. Thick legs and short, broad feet with hoofed toes (four per forefoot, three per hindfoot) supported its massive body. From its blunt snout jutted a thick Y-shaped horn, perhaps brandished at attacking creodonts. (Males might have used horns in jousting contests.) It lived in Oligocene North America, probably on open plains, crushing soft leaves between the big, square, low-crowned but enamel-hardened molars. Shoulder height: 2.5m (8ft).

2 **Moropus** resembled a big horse, but like all chalicotheres, had claws, low-crowned teeth unsuited for eating grass, and longer front limbs than hind limbs so that its back sloped down from shoulders to hips. Length: 3m (10ft). Time: Early–Mid Miocene. Place: North America.

Clash of titans (below)
As shown here, rival male titanotheres (brontotheres) might have locked horns like stags or banged heads like bighorn rams. Winning stags and rams rule herds of females; perhaps successful male titanotheres had harems, too.

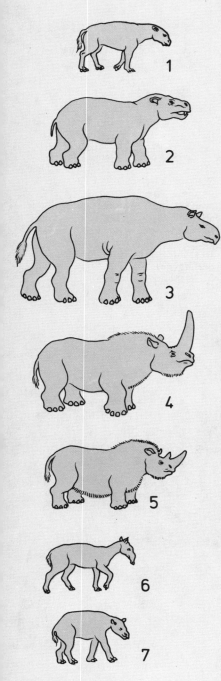

Tapirs and rhinoceroses

Tapirs and rhinoceroses appeared some 50 million years ago. Early kinds were small, agile beasts like early horses. But each of the two superfamilies had distinctive teeth for browsing. Rhinoceroses produced three families with scores of genera – some huge, many sprouting nasal horns. Tapirs remained small and primitive, with a heavy body, rounded back, short, stubby legs, and a flexible trunk-like snout for grasping forest plants. Several families evolved.

Rhinoceroses and tapirs reached all northern lands and Africa, and tapirs entered South America. But now rhinos only live in Africa and Asia; tapirs in Asia and South America.

1 **Hyracodon** was a small agile member of the hyracodontids, early "running" rhinos with long, slim legs, long three-toed feet, and a browser's molars. Length: 1.5m (5ft). Time: Oligocene. Place: North America.

2 **Metamynodon** was a hippo-like amynodontid rhino with short, thick limbs and short, broad feet. Time: Oligocene. Place: North America.

3 **Paraceratherium (Baluchitherium),** the largest-ever land mammal, was a rhinocerotid rhino. Long legs and longish neck helped its small hornless head browse giraffe-like among high branches. Shoulder height: 5.5m (18ft). Time: Oligocene–Miocene. Place: Asia.

4 **Elasmotherium,** an elephant-sized rhinocerotid, had a huge horn measuring almost 2m (6ft 6in). Time: Pleistocene. Place: Eurasia.

5. **Coelodonta,** the (rhinocerotid) woolly rhinoceros, survived Ice Age cold helped by its shaggy coat. Its head bore two horns – fore and aft. Time: Pleistocene. Place: Eurasia.

6 **Helaletes** was a small, early, agile helaletid tapir from Eocene North America and East Asia.

7 **Miotapirus** was a tapirid ancestor of modern tapirs, from Early Miocene North America.

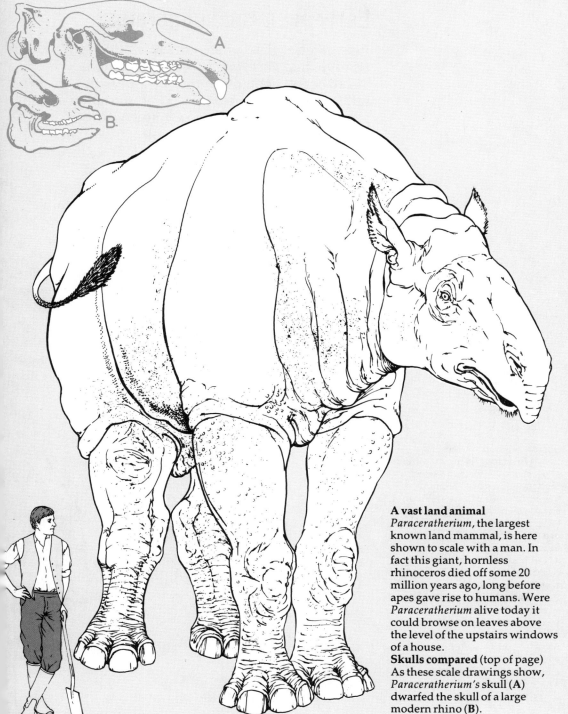

A vast land animal
Paraceratherium, the largest known land mammal, is here shown to scale with a man. In fact this giant, hornless rhinoceros died off some 20 million years ago, long before apes gave rise to humans. Were *Paraceratherium* alive today it could browse on leaves above the level of the upstairs windows of a house.

Skulls compared (top of page)
As these scale drawings show, *Paraceratherium's* skull (**A**) dwarfed the skull of a large modern rhino (**B**).

©DIAGRAM

Early even-toed ungulates

Pigs, camels, giraffes, sheep, cattle, and their relatives and ancestors make up the artiodactyls or "even-toed" hoofed mammals. Appearing over 50 million years ago, artiodactyls became today's major group of ungulates thanks largely to advances in digestion. We start by a brief look at two of their more primitive suborders. The extinct palaeodonts comprise two superfamilies: dichobunoids (ancestral artiodactyls) and entelodontoids (often called "giant pigs"). The suborder Suina contains three superfamilies: suoids (pigs and peccaries), anthracotheres (extinct pig-like beasts), and the hippopotamoids (hippopotamuses).

1 **Dichobune,** a small dichobunoid, had short limbs, four-toed feet, and low skull with long canine teeth and low-crowned molars. Time: Mid Eocene–Early Oligocene. Place: Europe.

2 **Archaeotherium** was a huge warthog-like entelodont. It had humped shoulders, thin legs, and a long, heavy head with lumpy cheeks. Probably it grubbed up roots. Height: 1m (3ft 3in). Time: Oligocene. Place: North America and East Asia.

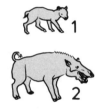

Sprinters and plodders
Bones from four living mammals show that odd-toed and even-toed ungulates of similar build have similar foot bones (shown shaded). These bones are longer in lightweight sprinters than in heavy plodders, and sprinters bear their weight on fewer toes.
a Foot bone of a horse, a speedy odd-toed ungulate that bears its weight on one toe.
b Foot bones of a chevrotain, a speedy even-toed ungulate that bears its weight on two toes.
c Foot bones of a rhinoceros, a plodding odd-toed ungulate that spreads its weight on three toes.
d Foot bones of a hippopotamus, a plodding even-toed ungulate that spreads its weight on four toes.

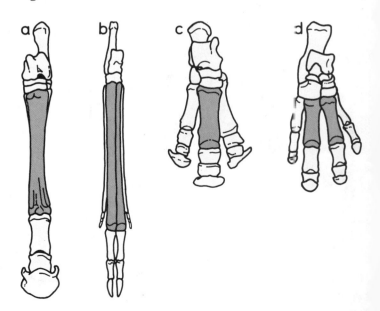

3 **Bothriodon,** an anthracothere, had a long body, short limbs with four-toed feet, and a long skull with 44 teeth including low-crowned molars. Length: 1.5m (5ft). Time: Late Eocene–Early Miocene. Place: Northern continents and Africa.

4 **Platygonus,** a peccary, had a short deep skull, big, shearing canine teeth, and long legs with reduced side toes. Length: 1m (3ft 3in). Time: Pliocene–Pleistocene. Place: North and South America.

5 **Hippopotamus,** the sole known genus in its superfamily, is a huge, heavy, aquatic mammal with short, thick, four-toed legs, tiny ears and eyes, and a vast mouth armed with tusk-like canine teeth. Length: up to 4.3m (14ft). Time: Mid Miocene to today. Place: Africa (once Eurasia too).

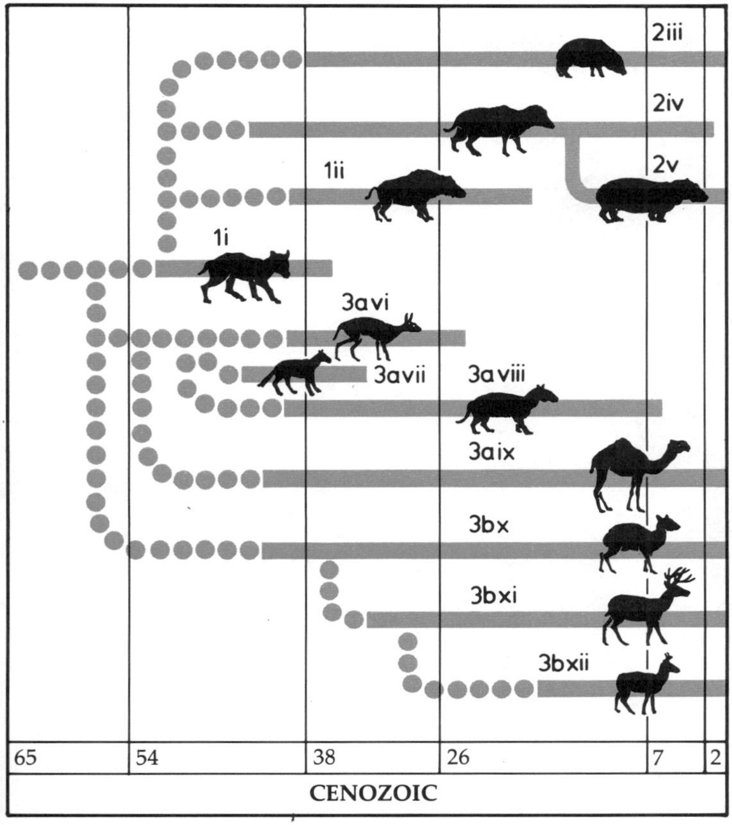

Artiodactyl family tree
This family tree shows the Artiodactyla ("even toed" ungulates), described on pages 186–195. We give suborders (1–3), infraorders (a–b), and superfamilies (i–xii).
1 Palaeodonta (palaeodonts)
i Dichobunoidea (ancestral artiodactyls)
ii Entelodontoidea (entelodonts)
2 Suina (pig-like artiodactyls)
iii Suoidea (pigs and peccaries)
iv Anthracotherioidea (anthracotheres)
v Hippopotamoidea (hippopotamuses)
3 Ruminantia (ruminants)
a Tylopoda (camels etc)
vi Cainotherioidea (cainotheres)
vii Anoplotheroidea (anoplotheres)
viii Merycoidodontoidea (oreodonts)
ix Cameloidea (camels and llamas)
b Pecora (pecorans)
x Traguloidea (tragulids etc)
xi Cervoidea (deer, giraffes, and kin)
xii Bovoidea (cattle, antelopes, prongbucks, etc)

The chart labels: 2iii, 2iv, 2v, 1ii, 1i, 3avi, 3avii, 3aviii, 3aix, 3bx, 3bxi, 3bxii

| 65 | 54 | 38 | 26 | 7 | 2 |

CENOZOIC

Camels and their kin

The most successful of all hoofed animals alive are those that ruminate, or "chew the cud". These so-called ruminants can snatch a meal of leaves, run away if set upon by carnivores, and then digest their meal at leisure. The suborder's more primitive members form the tylopods, an infraorder containing camels and their prehistoric kin. Their front teeth and cheek teeth differ less strikingly than those of pigs, and cheek teeth bear crescent-crested crowns. Tylopods mostly browse on trees and shrubs.

There were four superfamilies: small, rabbit-like cainotheres; primitive anoplotheres; heavily built, rather pig-like merycoidodonts (also often known as oreodonts); and cameloids, including camels and llamas. Tylopods appeared about 45 million years ago. Some kinds grew numerous and widespread. But only cameloids survive. They evolved in North

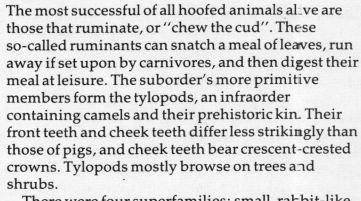

Giant and dwarf

Two prehistoric camels shown here to scale illustrate the range of sizes in their superfamily 10 million years ago.
A *Alticamelus* had a long neck and stilt-like limbs. This "giraffe camel" held its head 3m (10ft) high. It could have browsed on leafy twigs high off the ground.
B *Stenomylus* was tiny, and browsed on low vegetation. But it had long, slender limbs designed for sprinting. *Stenomylus* and its close kin were the camel counterparts of gazelles.

America, reached South America, Eurasia, and Africa, but then died out in North America, Europe, and Africa.

1 Cainotherium was a rabbit-sized cainothere. Big eyes and keen ears warned of danger, and it escaped by bounding off on long hind limbs. It lived in Mid Oligocene–Mid Miocene Europe.

2 Anoplotherium belonged to the anoplotheres: heavily built, tapir-sized beasts with clawed toes and long tails. Shoulder height: 1m (3ft 3in). Time: Late Eocene–Early Oligocene. Place: Europe.

3 Merycoidodon was a sheep-sized oreodont, with rather pig-like proportions: large head, short neck, long body, and short limbs. Time: Early–Mid Oligocene. Place: North America.

4 Agrichoerus belonged to the agrichoeres – a family of slim oreodonts with long head, body, and tail, and clawed feet used perhaps in climbing trees or digging up edible roots. This sheep-sized beast lived in Oligocene–Early Miocene North America.

5 Poëbrotherium was a sheep-sized early camel with short legs, two-toed feet, and a full set of teeth. It looked like a tiny llama. Time: Oligocene. Place: North America.

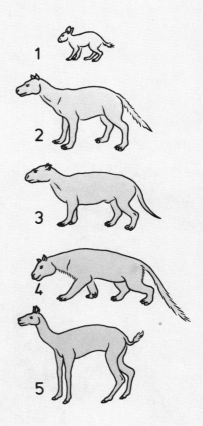

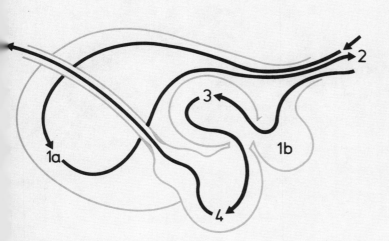

Ruminant stomach
This section diagram shows what happens to swallowed food inside a ruminant's four-chambered stomach.
1a Most swallowed food enters the rumen, the largest chamber; digestion starts here.
1b Tough plant fibres and stones go to the reticulum.
2 Pulped food returns to the mouth for extra crushing.
3 Re-swallowed food has water squeezed out in the omasum.
4 Enzymes extract proteins from food in the abomasum.

Primitive pecorans

Cattle, sheep, deer, and close relatives make up the pecorans – a more advanced cud-chewing infraorder than the tylopods (camel and their kin). Pecorans have very complex stomachs. Most are graceful, with long, slim limbs. Their main defence is speed, though most have horns or antlers, or long, sharp canine teeth. They crop leaves by pressing them between the lower front teeth and an upper horny pad; the top front teeth are missing.

Pecorans first appeared over 40 million years ago, seemingly in Asia, and spread to other continents. Three superfamilies evolved. First we look at the traguloids: primitive mostly prehistoric pecorans including small hornless deer-like beasts with less complicated stomachs than true deer. Our examples come from different families.

Horned traguloid
Syndyoceras from Miocene North America belonged to the protoceratids, traguloids of which the males developed spectacularly strange horns. It included these features.
a Small deer size
b A pair of curved horns growing above the eyes
c A diverging pair of horns forming a prong above the nose

d Long, slim limbs
e Only the third and fourth toes functional
f Metacarpal bones on front feet not fused as in higher ruminants

190

1 **Archaeomeryx** was no bigger than a large rabbit. It had a curved back and longer hind limbs than front limbs. It looked like living mouse deer (chevrotains), but belonged to the primitive hypertragulids which had a long, canine-like premolar tooth in each lower jaw. Time: Late Eocene. Place: East Asia.

2 **Synthetoceras** belonged to the protoceratids, deer-sized traguloids whose males grew horns. *Synthetoceras* males had a Y-shaped horn jutting up and forward from the nose and shorter horns curving back and up behind the eyes. Time: Early–Late Miocene. Place: North America.

3 **Tragulus,** the Asian chevrotain, is one of only two survivors of the traguloids. Like others in its family (the tragulids), this looks like a rodent about 30cm (1ft) high. It lacks horns but has long, tusk-like upper canines, and four toes per foot (the outer two are short and useless). Time: Late Pliocene to today. Place: South-East Asian forests.

Leptomeryx skull
This life size skull of an early pecoran has the following features.
A No horns
B Sharp upper canine teeth
C Toothless cropping pad
D Cropping front teeth
E Gap developing between front and back teeth

© DIAGRAM

Deer and giraffes

The cervoids (deer, giraffes, and relatives) and bovoids (cattle and their kin) are more "progressive" ruminants than their ancestors the traguloids. Mostly their upper canine teeth are short or missing, but their heads sprout horns or antlers. Certain bones inside their long, two-toed limbs have shrunk or almost vanished. And living members of these groups have a four-chambered stomach, better able to digest tough plant foods than the three-chambered stomach of the chevrotains.

Members of the cervoid superfamily live mostly in wooded countryside. Their low-crowned cheek teeth are designed to browse on trees or shrubs. Male deer grow and shed antlers each year; giraffes grow permanent skin-covered horns. We show five prehistoric examples from the three cervoid families, all established 25 million years ago.

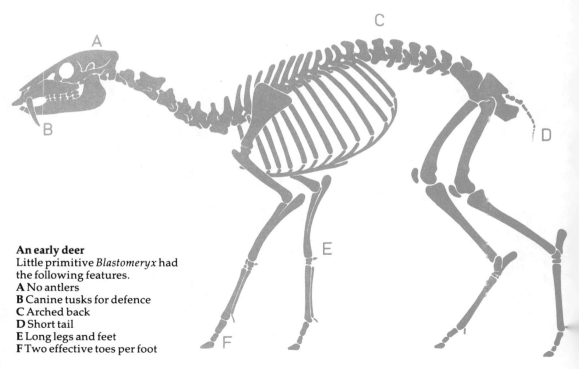

An early deer
Little primitive *Blastomeryx* had the following features.
A No antlers
B Canine tusks for defence
C Arched back
D Short tail
E Long legs and feet
F Two effective toes per foot

1 **Blastomeryx** had "advanced" limbs but old-fashioned tusk-like upper canines and no horns. It belonged to the early, primitive palaeomerycid family. Length: 76cm (30in). Time: Miocene. Place: North America.

2 **Cranioceras,** a North American palaeomerycid, had tall, pronged horns and a backswept third horn. Length: 1.5m (5ft). Time: Miocene–Pliocene.

3 **Megaloceros,** misnamed the "Irish elk", was a giant relative of the fallow deer, with antlers 3m (10ft) across. It lived in Pleistocene Eurasia and died off about 2500 years ago.

4 **Palaeotragus,** a giraffid ancestor of okapis and giraffes, resembled the okapi. Time: Miocene. Place: Eurasia, Africa.

5 **Sivatherium** was a large giraffid with a big head, but shorter legs and neck than a giraffe. Males grew two pairs of horns, one long and branched. Shoulder height: 2.2m (7ft). Time: Pleistocene. Place: South Asia and Africa.

The antler cycle
Like living deer, *Megaloceros* would have grown and shed its antlers every year. The annual cycle went like this.
A New antlers sprout in summer, nourished and protected by a covering called velvet.

B Bony antlers reach their full extent in autumn, when rival stags perhaps locked them in combat.
C Antlers shed in spring reveal the pedicle or base from which next season's antlers grow.

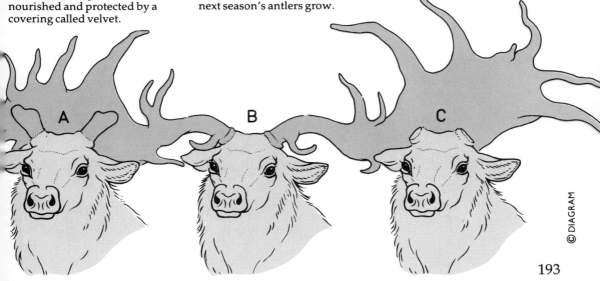

Cattle and their kin

The top grass eaters in the world today are bovoids – members of the cattle superfamily. Cattle, sheep, goats, antelopes, musk-oxen, and prongbucks between them share out almost all the grasslands of the world, with horses as their only large competitors. The bovoids' teeth, stomachs, legs, and feet superbly suit them for a life as grazers on the plains. Both sexes grow strong horns they never shed – living spears on which they can impale attacking carnivores.

Bovoids seemingly evolved from tragulids some 15 million years ago, later than giraffes or deer. Five million years ago great herds were spreading through the north. In time they swarmed on grassy slopes and plains across Eurasia, Africa, and North America. Experts disagree about how to subdivide bovoids. Our examples show five contrasting types.

1 **Gazella** is a small, delicately built antelope, a desert and savanna sprinter. Shoulder height: 65–100cm (26–40in). Time: Late Miocene onward. Place: Africa, Asia, and once also Europe.

Unusual horns (left)
Unlike other bovoids, prongbucks grow forked horns and annually shed the outer sheath, seen here in section.
a Permanent bony horn core
b Hair covering the core
c Horn formed by outgrowth of the hair – growth takes four months from when the old horn has been shed

Prongbuck duellists (below)
Rival *Merycodus* males could have locked their long forked horns in ritual combat, as shown here. As in most bovoid duels, such fights were probably trials of strength, not combats to the death. The weaker creature would have given up and run away.

2 **Mesembriportax** was a large, rather heavy-bodied antelope related to the living nilgai. It had long legs and uniquely forked horns. Time: Early Pliocene. Place: South Africa.

3 **Bison** The long-horned species shown lived in Late Pleistocene North America. Shoulder height: up to 1.8m (6ft).

4 **Myotragus,** the Balearic cave goat, lived on West Mediterranean islands in Pleistocene and Recent times. Its large lower middle incisor teeth resembled rodents'. Shoulder height: 50cm (20in).

5 **Merycodus** was a small, early, deer-like prongbuck "antelope" with long forked horns. Prongbucks keep the bony horn cores but annually shed the horny sheaths. Time: Mid Miocene–Late Miocene. Other prongbucks (some bizarrely horned) evolved, all in North America. Only one survives.

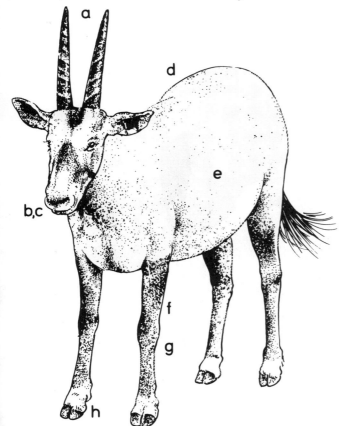

Bovoid features (left)
Palaeoreas was a small Miocene ruminant of Europe, Asia, and North Africa. Its spiral horns foreshadowed those of the big modern antelopes called oryxes. *Palaeoreas* and its bovoid kin owed their success to features such as the following.
a Both sexes permanently horned
b Wear-resistant teeth, with high crowns strengthened by folded ridges of enamel
c Lower incisors cropping against a hard toothless pad
d Sturdy body
e Four-chambered stomach for digesting tough leaves
f Long limbs for running
g Fused cannon bone reducing risk of sprains
h Two-toed "cloven" hooves

©DIAGRAM

"Toothless" mammals

Strange mammals with few teeth make up the order Edentata. Edentates evolved in the Americas while these were isolated. They include the living anteaters, sloths, and armadillos, and their extinct relatives the huge, astonishingly armoured glyptodonts and unwieldy ground sloths. Most edentates lack front teeth and have a few, simple, rootless cheek teeth. The brain is small. The limbs are short, strong, and tipped with long, curved claws. All edentates were and are slow-moving eaters of plants, carrion, or insects.

Perhaps the first-known kinds were the so-called palaeanodonts from the North America of 60 million years ago. But maybe those gave rise to another order (Pholidota): the scaly Old World pangolins. Most fossil edentates arose in what was then the island continent of South America, though a land-bridge later let some into North America. Local tales and finds of hairy hides hint that ground sloths survived in southern Argentina until four centuries ago. Quite likely man killed off these last big edentates.

Glyptodon's body plan
a Deep, short skull with a small brain case
b Teeth only in sides of jaw
c Each tooth like three short pillars stuck together in a row, and without enamel
d Many vertebrae fused to help support the shell
e Short, massive limb bones
f Broad shoulder girdle
g Massive hip bones
h Heavy, bony tail
i Five-toed feet with hoof-like claws
j Solid armour (evolved from bony studs and plates set in the skin)

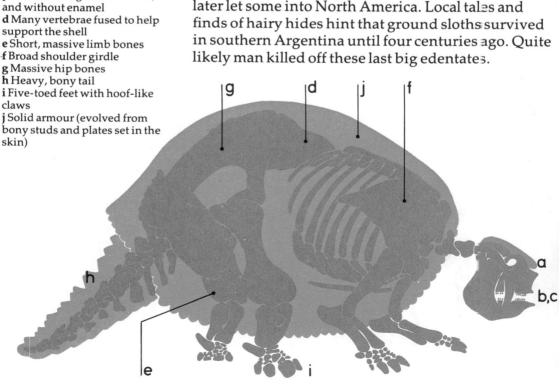

These three examples show a palaeanodont and two enormous fossil edentates.

1 Metacheiromys was a palaeanodont with a long, low head, sharp canine teeth, but maybe horny pads instead of cheek teeth. It had short legs, sharp claws, and a long, heavy tail. Length: 45cm (18in). Time: Middle Eocene (about 48 million years ago). Place: North America.

2 Glyptodon was a huge "mammal tortoise"—a glyptodont descended from early armadillos. Horny sheaths covered the many bony plates that formed a great dome-shaped shell around its body. A bony cap crowned its skull, and bony rings armoured a thick, heavy tail, swung to fend off enemies. Length: about 3m (10ft). Time: Pliocene–Pleistocene. Place: South America.

3 Megatherium, the largest ground sloth, was an elephantine tree-top browser. Length: 6m (20ft). Time: Late Pliocene–Pleistocene. Place: South America to south-east USA.

Megatherium's life style
Megatherium is here shown to scale with a man. The monster walked on its knuckles and the sides of its feet. Females might have carried a baby on their back. A strong tail propped up *Megatherium* as it reared to claw leaves into its mouth and chewed them with its peg-like teeth.

197

Whales

Whales are mammals so well adapted for water life that they swim as easily as fishes and cannot walk on land. Fossil clues suggest that their order (Cetacea) evolved from mesonychids – a flesh-eating group of condylarths, the first hoofed mammals. By 52 million years ago these probably gave rise to *Pakicetus*, known from a skull unearthed in Pakistan and described in 1983. Sharing features found in whales and tapirs, *Pakicetus* was about 1.8m (6ft) long, lived near water, but could not dive deep or hear well when submerged. By 40 million years ago such beasts gave rise to three suborders of true whales worldwide. Buoyed up by water, some evolved bigger bodies than the largest-ever land mammal.

A snaky whale
Here the big early archaeocete *Basilosaurus*, also known as *Zeuglodon*, is shown in pursuit of herring. At over 20m (66ft), this monster rivalled many a large modern whale for length, but it had a slender rather snake-like body. The saw-edged teeth were suitable for seizing fishes. *Basilosaurus* fossils crop up in rocks of what were ancient sea beds as far apart as Africa and North America.

1 **Prozeuglodon,** found in Mid-Late Eocene North Africa, grew maybe 3m (10ft) long. It had a long snout, peg-like front teeth, and three-ridged saw-like cheek teeth. It belonged to the archaeocetes, a primitive Eocene–Late Oligocene suborder.

2 **Prosqualodon,** a more advanced small whale, had its blowhole above and behind the eyes. But its triangular shark-like cheek teeth were "old fashioned". It resembled dolphins and, like those, belonged to the odontocetes (toothed whales). Length: 2.3m (7ft 6in). Time: Oligocene–Early Miocene. Place: southern oceans.

3 **Cetotherium** probably resembled a small version of the living grey whale – a mysticete (baleen whale). These toothless whales trap swarms of shrimp-like creatures on the fringed baleen plates that hang from their upper jaws. Length: perhaps 4m (13ft). Time: Mid–Late Miocene. Place: Europe.

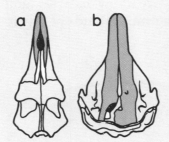
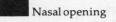
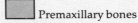

Back to the blowhole
Skulls of prehistoric whales show the nasal opening moving back to become a blowhole.
a *Prozeuglodon,* an Eocene whale from North Africa
b *Aulophyseter,* a Miocene whale from North America

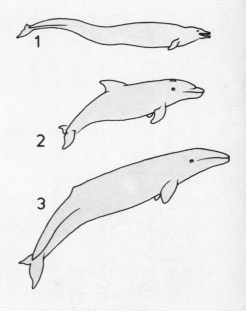

■ Nasal opening

▒ Premaxillary bones

□ Nasal bones

Rodents and rabbits

In numbers, variety, and distribution, the most successful mammals ever have been members of the rodent order: squirrels, rats, mole rats, cavies, beavers, porcupines, and many more. Between them rodents live in trees, on mountains, underground, in streams and swamps – everywhere from polar wastes to steamy forests in the tropics.

Small size, fast breeding, and ability to gnaw and digest foods as hard as wood contributed to their success. This started when an insectivore-type ancestor gave rise to the first squirrel-like rodent more than 60 million years ago. From northern continents, that pioneer's descendants reached every continent except Antarctica.

Soon after rodents came the rodent-like lagomorphs: rabbits, hares, and pikas. Lagomorphs have eight long chisel-like incisor teeth (rodents have just four) and there are other differences. What we know of early lagomorphs and rodents owes much to finds of tiny fossil teeth.

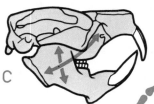

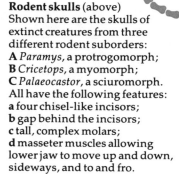

Rodent skeleton
This *Paramys* skeleton shows the following rodent features.
e Long, low skull
f Flexible fore limbs
g Strong hind limbs
h Five clawed toes per limb

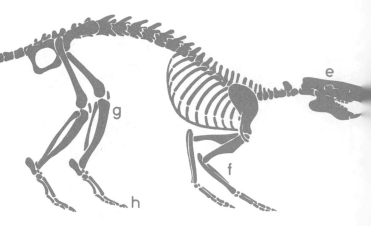

Rodent skulls (above)
Shown here are the skulls of extinct creatures from three different rodent suborders:
A *Paramys*, a protrogomorph;
B *Cricetops*, a myomorph;
C *Palaeocastor*, a sciuromorph.
All have the following features:
a four chisel-like incisors;
b gap behind the incisors;
c tall, complex molars;
d masseter muscles allowing lower jaw to move up and down, sideways, and to and fro.

1 **Paramys,** the first-known rodent, was a squirrel-like climber in the primitive suborder protrogomorphs. Length 60cm (2ft). Time: Late Palaeocene–Mid Eocene. Place: North America and Europe.

2 **Epigaulus,** a two-horned gopher, was a burrowing protrogomorph rodent. Its horns served in defence or maybe as a pair of shovels. Length: 26cm (10in). Time: Miocene. Place: North America.

3 **Sciurus,** the squirrel genus found today in North America and Europe, is a "living fossil" in the rodent suborder Sciuromorpha; its history dates back 38 million years.

4 **Castoroides** was a land-based beaver almost as big as a black bear. Length: 2.3m (7ft 6in). Time: Pleistocene. Place: North America.

5 **Eocardia** was a Miocene guinea-pig-like member of the cavioids, a South-American rodent suborder. (*Eumegamys,* an almost hippo-sized Pliocene cavioid, was the largest ever rodent.)

6 **Lepus,** the modern hare, evolved five million years or so ago. Place: northern continents and Africa. Order: Lagomorpha.

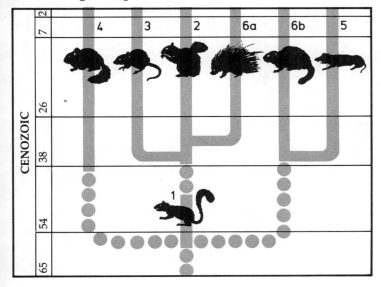

Rodent family tree
Zoologists group most of the 45 or so rodent families into suborders based on cheek teeth and jaw muscle design. We show the five suborders (**1–5**) and two well-known families unassigned to any suborder (**6**).
1 Protrogomorpha (primitive rodents)
2 Sciuromorpha (squirrels)
3 Myomorpha (rats, mice, etc – the largest rodent group)
4 Caviomorpha (South American rodents such as guinea pigs and chinchillas)
5 Phiomorpha (mole rats etc)
6 Unassigned to a suborder:
a Hystricidae (Old World porcupines)
b Castoridae (beavers)

©DIAGRAM

201

Chapter 9

RECORDS IN THE ROCKS

Immense changes have transformed the surface of the Earth since its creation. Prolonged rains filled the ocean basins. Later, continents collided and broke up. Their shifting created mountains and altered climates: deserts grew and shrank; hot lands cooled down and vice versa. Rarely, huge rocks hurtled in from space and punched holes in the Earth's crust, sometimes perhaps explosively enough to alter climates suddenly.

Living things responded to such changes with waves of extinction followed by bursts of evolution. All this helped to give the phases of Earth's history their special characters.

The next pages draw a thumbnail sketch of life through the periods and epochs of our ever-changing planet.

Eighteenth-century antiquaries examine rock layers crammed with ancient sea shells. Fossils in rocks laid down at different times allow palaeontologists to reconstruct past life period by period, epoch by epoch. (Illustration from *The History and Antiquities of Harwich and Dovercourt* by Samuel Dale.)

Precambrian time

Precambrian time – time before the Cambrian Period began – occupied Earth's first 4000 million years – 87 per cent of our planet's past. Rocks laid down then still largely form the cores of continents. Some hold the fossil clues that show when life began and early living things evolved. Microscopic one-celled life forms appeared in seas at least 3500 million years ago. By 3200 million years ago, blue-green algae had begun enriching the atmosphere with oxygen. In time they made life possible for more complex living things. Soft-bodied, many-celled water animals were burrowing through underwater mud 1000 million years ago. By 680 million years ago, soft corals, jellyfish, and worms all flourished off a sandy shore in South Australia. Similar creatures also thrived in what today are England and Newfoundland.

Precambrian world
This world map shows land masses in Late Precambrian time. Lines represent Equator, Tropics, and Polar Circles.

Precambrian life
These fossil organisms lived in Precambrian time. They are not shown to scale.

MONERA
a *Kakabekia*

PLANTS
b Green alga

INVERTEBRATES
c Sponge
d Annelid worms
e Coelenterates

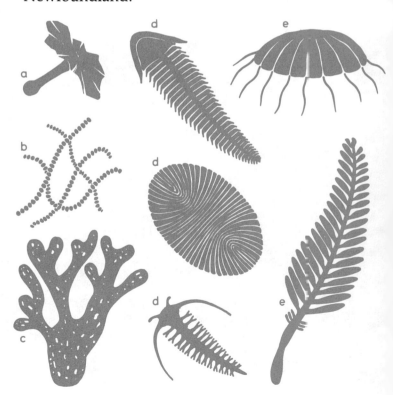

Cambrian Period

Fossils suddenly appear more numerous and varied in Cambrian times (600–500 million years ago), the dawn of the Palaeozoic Era ("age of ancient life"). Scientists first studied them in Wales (called Cambria in Latin). But Cambrian rocks occur in North America and other continents as well as Europe. The fine-grained Burgess Shales of south-west Canada preserved a rich sample of life below the waves 550 million years ago. Down here lived jellyfishes, sponges, starfishes, worms, and velvet worms – all relatives of animals alive today. Most striking, though, were creatures that had gained the knack of building hard protective shells from chemicals dissolved in water. The commonest were trilobites, many of them sea-bed scavengers. There was even a little lancelet-like creature – forerunner of the jawless fishes which appeared just before the Cambrian Period closed.

Cambrian world
This world map shows land masses in Cambrian time. Lines represent Equator, Tropics, and Polar Circles.

Cambrian life
These fossil organisms lived in Cambrian time. They are not shown to scale.

       INVERTEBRATES
a Annelid worm
b Molluscs
c Coelenterate
d Arthropods
e Lancelet

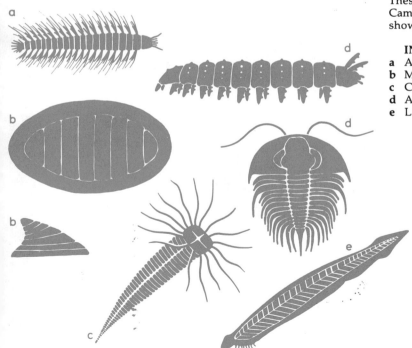

©DIAGRAM

Ordovician world
This world map shows land masses in Ordovician time. Lines represent Equator, Tropics, and Polar Circles.

Ordovician life
These fossil organisms lived in Ordovician time. They are not shown to scale.

INVERTEBRATES
a Coelenterate
b Molluscs
c Arthropod
d Tentaculate
e Echinoderms
f Branchiotreme

Ordovician Period

The Ordovician Period (500–440 million years ago) is named after the Ordovices, an ancient Celtic tribe of western Wales, where scientists first studied Ordovician fossils. In Ordovician times, northern landmasses were coming together, and southern continents already formed a single mass of land. The South Pole lay over North Africa, much of which was under ice. Shallow seas repeatedly invaded North America.

Ordovician fossils show that animals and plants still lived only in the sea. Many resembled Cambrian ancestors. Trilobites were now most numerous. Graptolites and lampshells teemed below the waves. Molluscs evolved apace – bivalves resembling modern clams and oysters, and gastropods (one-shelled molluscs such as whelks and limpets). Now, too, those early vertebrates the jawless fishes increased in number.

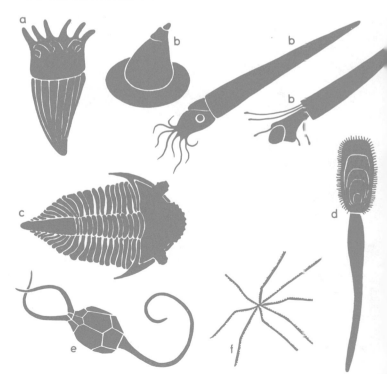

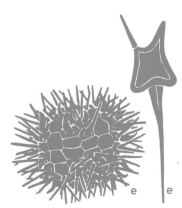

Silurian Period

The Silurian Period (440–395 million years ago) is named after the Silures, an ancient tribe astride the Welsh–English border. Silurian fossils occur in all landmasses but Antarctica.

By the end of Silurian time, colliding continents had raised up mountains and forged two supercontinents: Laurasia in the north, and Gondwanaland in the south. Life thrived largely in warm shallow seas that invaded much of Laurasia. The first jawed fishes (acanthodians and placoderms) appeared, some of them doubtless hunted by the giant "sea scorpions". Solitary corals built great reefs, and shallow sea floors supported a rich array of sea lilies, lampshells, corals, trilobites, graptolites, and molluscs. Jawless fishes invaded lakes and rivers. The first land plants appeared, followed soon by creatures such as scorpions and millipedes.

Silurian world
This world map shows land masses in Silurian time. Lines represent Equator, Tropics, and Polar Circles.

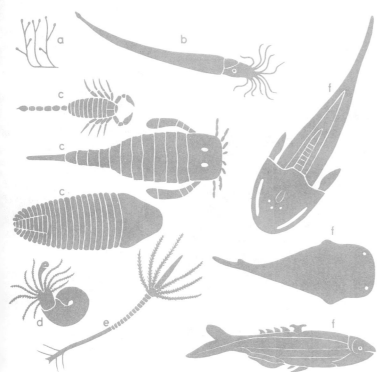

Silurian life
These fossil organisms lived in Silurian time. A re-dating of this period may put some items later. They are not shown to scale.

    PLANTS
a Rhyniophyte

    INVERTEBRATES
b Mollusc
c Arthropods
d Annelid worm
e Echinoderm

    FISHES
f Agnathans

©DIAGRAM

Devonian Period

The Devonian Period (395–345 million years ago) is named from shales, slates, and Old Red Sandstone laid down in Devon, England. But such rocks occur in every continent. During Devonian times Laurasia and Gondwanaland drew closer together until they met at what are now North and South America to form a single supercontinent: Pangaea ("all Earth"). Shallow sea invaded Laurasia, and mountains rose where North America and Europe had collided. Climates everywhere were warm.

The Devonian is aptly called the Age of Fishes, for now these backboned animals diversified and multiplied enormously. Jawless fishes shared the waters with the more progressive jawed fishes destined to replace them. Among these were huge placoderms, early sharks, and early bony fishes,

Devonian world
This world map shows land masses in Devonian time. Lines represent Equator, Tropics, and Polar Circles.

Devonian life
These fossil organisms represent some of those that flourished in Devonian time. They are not shown to scale.

PLANTS
a *Asteroxylon*
b Fern ancestor

INVERTEBRATES
c Mollusc
d Arthropod

FISHES
e Agnathan
f Placoderms
g Acanthodians
h Chondrichthyan
i Sarcopterygian

AMPHIBIANS
j Labyrinthodont

including the fleshy-finned ancestors of amphibians. In seas, vast coral reefs produced immense tracts of limy rock, and the ancestors of ammonites appeared. But trilobites and graptolites grew scarcer.

By Early Devonian times, there were many land plants equipped with tubes for sucking moisture from the soil. Scale trees, giant horsetails, and feathery tree ferns produced the world's first forests. These mostly thrived on swampy lands, but the first seed-bearing plants foreshadowed kinds that later colonized dry land. Greenland's swampy tropical forests became home to the first, low-slung amphibians. Here, too, crawled spider-like arachnids and the first wingless insects – bristletails and springtails.

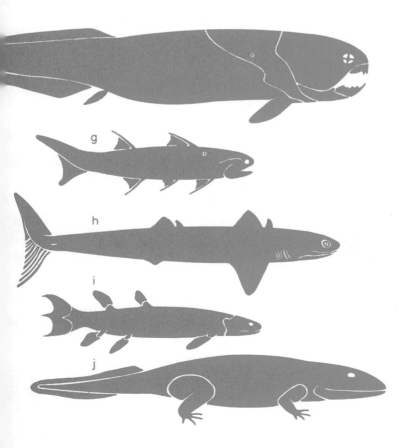

Carboniferous Period

The Carboniferous Period (345–280 million years ago) gets its name from thick bands of carbon in the form of coal laid down when shallow seas drowned tropical forests in what are now North America, Europe, and elsewhere.

The period had two distinct halves, with different names in the Americas. Mississippian, or Early Carboniferous, rocks (345–300 million years old) include limestones formed when limy mucs were laid down in a shallow sea in what is now the Mississippi Valley. Arthropods, bryozoans, crinoids, corals, and molluscs flourished in these waters.

Pennsylvanian, or Late Carboniferous, rocks formed 300–280 million years ago. They include Pennsylvania's coal measures, produced when shallow sea drowned tropical forests that covered much of lowland North America. The rich variety of

Carboniferous world
This world map shows land masses in Carboniferous time. Lines represent Equator, Tropics, and Polar Circles.

Carboniferous life
These fossil organisms represent some of those that flourished in Carboniferous time. They are not shown to scale.

PLANTS
a Horsetail
b Gymnosperm
c Club moss

INVERTEBRATES
d Arthropods

FISHES
e Chondrichthyan
f Acanthodian

AMPHIBIANS
g Labyrinthodont

REPTILES
h Cotylosaur

life in Pennsylvanian coal swamps helps to earn the
Carboniferous Period its nickname "Age of
Amphibians". Sprawling amphibians – some as big
as crocodiles – lurked on mudbanks below the tall
scale trees, tree ferns, and giant horsetails. Others
hunted fishes in the pools they shared with
lungfishes. Amphibians with sturdy limbs walked
easily on land. Those with tiny limbs or none swam
like eels or burrowed through leaf litter on the forest
floor. Here snails slithered through the humid
undergrowth. Rotting logs sheltered centipedes,
scorpions, and countless kinds of cockroach. Old tree
stumps were the dens of small, early, insect-eating
reptiles. Above flew or hovered giant "dragonflies"
and lesser insects. Meanwhile, vast ice sheets
smothered much of the world's great southern
landmass.

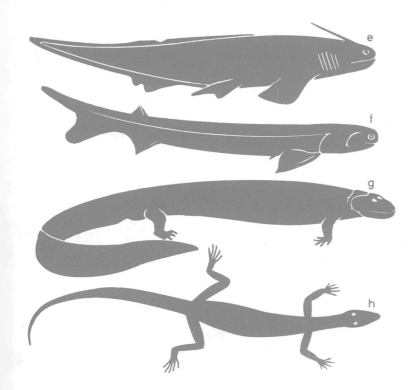

Permian Period

Permian world
This world map shows land masses in Permian time. Lines represent Equator, Tropics, and Polar Circles.

Fossil-rich rocks near Perm in Russia explain the name given to the Permian Period (280–225 million years ago). But much of what we know of Permian life on land comes from the Red Beds of Texas and South Africa. As the supercontinent Pangaea drifted north, glaciers retreated southwards. Great Earth movements threw up mountain ranges ancestral to the Appalachians, Rockies, Alps, and Urals. Fossil plants betray distinct climatic zones, and a trend to seasonally cooler, drier climates. Conifers began replacing scale trees and giant horsetails. The new forests largely featured firs and pines.

Now, reptiles took over from amphibians to dominate life on land. First came cotylosaurs, then mammal-like reptiles – primitive pelycosaurs and the more advanced therapsids. Now, too, the lizards' ancestors were scampering around, and sturdy

Permian life
These fossil organisms represent some of those that flourished in Permian time. They are not shown to scale.

PLANTS
a Tree fern
b Conifer

INVERTEBRATES
c Arthropods

FISHES
d Chondrichthyan
e Bony fish

AMPHIBIANS
f Labyrinthodonts
g Lepospondyls

REPTILES
h Cotylosaurs
i Mesosaur
j Therapsid
k Pelycosaur
l Eosuchian

amphibians like *Cacops* and *Diadectes* lumbered overland. But the last anthracosaur and lepospondyl amphibians died out. New major insect groups – beetles, bugs, and cicadas – were emerging.

Changes also happened in the waters. The last acanthodians died out, along with rhipidistians, the fleshy-finned fishes that had given rise to amphibians. But bony fishes teemed in lakes and rivers, and some found their way into the sea. Here, ammonoids were plentiful, but trilobites and sea scorpions became extinct at last. In fact, dozens of major groups of creatures (most marine invertebrates) vanished from the fossil record as the Permian Period ended. This mass death marks the ending of the Palaeozoic Era – the "age of ancient life".

Triassic world
This world map shows land masses in Triassic time. Lines represent Equator, Tropics, and Polar Circles.

Triassic life
These fossil organisms represent some of those that flourished in Triassic time. They are not shown to scale.

PLANTS
a Gymnosperm

INVERTEBRATES
b Arthropod
c Molluscs

AMPHIBIANS
d Anuran
e Labyrinthodonts

REPTILES
f Euryapsid
g Euryapsid or eosuchian
h Chelonian
i Ichthyosaur
j Cotylosaur
k Crocodilian
l Therapsids
m Rhynchocephalian
n Thecodont
o Saurischian dinosaur
p Ornithischian dinosaur

MAMMALS
q Eotherian

Triassic Period

The Triassic Period (225–193 million years ago) marks the start of the Mesozoic Era or "age of middle life", also called the Age of Dinosaurs. The Triassic gets its name from the Latin *trias* ("three") after three rock layers dating from this period and found in Germany.

In Triassic times, the vast supercontinent Pangaea was splitting up to reform the northern and southern units, Laurasia and Gondwanaland. Land animals could still colonize the world by simply walking overland, but deserts occupied vast inland tracts. Conifers and other plants designed for cool or dry conditions thrived, at the expense of moisture-loving kinds, though ferns and horsetails flourished by the waterside. Now, too, there were palm-like cycads and bennettitaleans, also yews and ginkgoes. This was indeed an age of reptiles. Early on there was

a wealth of rhynchosaurs, thecodonts, and mammal-like reptiles. But all these faded out, to be replaced by others. By Mid-Late Triassic times, early saurischian dinosaurs shared the lands with tortoises and lizards, and maybe with the earliest ornithischian dinosaurs and shrew-like mammals. Above their heads flapped skin-winged pterosaurs. Lakes and rivers formed the homes of crocodiles, while early frogs hopped amid damp herbage at the water's rim. Shallow seas afforded hunting grounds for strange marine reptiles: placodonts, nothosaurs, and those uncannily dolphin-like creatures, the flippered ichthyosaurs. Dinosaurs, crocodiles, and pterosaurs – the masters on land, in fresh water, and in air – were all members of one group: the archosaurs. These ruling reptiles dominated life on land all through the rest of Mesozoic time.

©DIAGRAM

Jurassic Period

The Jurassic Period (193–136 million years ago) is named from rocks formed at this time in the Jura Mountains of France and Switzerland. Gondwanaland began breaking up; the North Atlantic Ocean opened and shallow seas invaded parts of Central North America and Europe. But land animals could still walk freely everywhere. Climates now were largely warm and moist, and rains watered what had been Triassic deserts.

Palm-like bennettitaleans and cycads, along with ferns and tree ferns, flourished thickly on moist river banks. A zoo of plant-eating dinosaurs browsed on this vegetation. Early armoured dinosaurs probably munched low-growing ferns and fungi. Knee-high leaves provided food for agile ornithopods. Stegosaurs might have reared on hind limbs to browse on low tree branches. But the "giraffes" of the Jurassic would have been huge sauropods

Jurassic world
This world map shows land masses in Jurassic time. Lines represent Equator, Tropics, and Polar Circles.

Jurassic life
These fossil organisms represent some of those that flourished in Jurassic time. They are not shown to scale.

PLANTS
a Conifer
b Bennettitalean

INVERTEBRATES
c Tentaculate
d Arthropod
e Molluscs

FISHES
f Bony fish
g Chondrichthyan

AMPHIBIANS
h Salamander

REPTILES
i Plesiosaur
j Ichthyosaur
k Crocodilian
l Pterosaur
m Ornithischian dinosaur
n Saurischian dinosaurs

BIRDS
o *Archaeopteryx*

capable of cropping treetop leaves beyond the reach
of other dinosaurs. These monsters beat paths
through forests, making gaps where smaller
dinosaurs could follow and find food. Sauropod
droppings fed the soil, and nourished seedlings that
in time became large trees. But there were hunting
dinosaurs as well. Small sprinters like *Coelurus*
chased lizards through the undergrowth or
scavenged at the kills of beasts like *Allosaurus* – a
great flesh-eating carnosaur that could have preyed
upon the sauropods. Now early birds took to the air,
joining strange furry-bodied pterosaurs. There was
also a crawling, buzzing menagerie of brand new
insects, ancestral to the living earwigs, flies, caddis
flies, bees, wasps, and ants. Meanwhile long-necked
and short-necked plesiosaurs, and ichthyosaurs
swam through the shallow seas in search of fish or
ammonites.

Cretaceous Period

The Cretaceous Period (136–65 million years ago) owes its name to vast thicknesses of chalk laid down in shallow seas (*creta* is the Latin name for chalk). By Late Cretaceous times, the break-up of the supercontinents Gondwanaland and Laurasia was well advanced, and continents were taking on their modern outlines and positions. Southern continents became vast islands. Seas split the northern landmass Laurasia in two: Asiamerica (East Asia with Western North America) and Euramerica (Europe with eastern North America). Mountain building began pushing up the present Rockies, Andes, and other mighty groups of mountains.

These changes isolated some groups of dinosaurs, so kinds that evolved in Asiamerica could no longer easily reach other continents, and vice versa. Meanwhile climates tended to grow cooler. Flowering plants evolved and spread explosively during the Cretaceous Period, along with the pollinating bees and butterflies. By 70 million years ago hickories, magnolias, and oaks formed streamside forests in what is now Alberta, with china firs, giant sequoias, and swamp cypresses covering more

Cretaceous world
This world map shows land masses in Cretaceous time. Lines represent Equator, Tropics, and Polar Circles.

Cretaceous life
These fossil organisms represent some of those that flourished in Cretaceous time. They are not shown to scale.

PLANTS
a Flowering plant

INVERTEBRATES
b Molluscs
c Arthropod
d Echinoderm

FISHES
e Chondrichthyan
f Sarcopterygian

REPTILES
g Chelonians
h Plesiosaurs
i Eosuchian
j Squamata
k Crocodilian
l Pterosaur
m Saurischian dinosaurs
n Ornithischian dinosaurs

BIRDS
o *Ichthyornis*
p *Hesperornis*

MAMMALS
q Insectivore

218

swampy tracts. Duckbilled dinosaurs evolved powerful batteries of grinding teeth to chew new tough-leaved kinds of vegetation. Ferns and cycad cones probably provided food for the horned dinosaur *Anchiceratops*. Higher, drier land away from swamps was the likely home of boneheaded dinosaurs like *Pachycephalosaurus*. Toothless ostrich dinosaurs roamed lowland clearings. *Tyrannosaurus* and its kin probably killed off enough duckbilled dinosaurs to stop these damaging the forests by overbrowsing. Other creatures living in or near this swampy delta region included such "modern" forms as frogs, salamanders, softshell turtles, snakes, gulls, waders, and opossums. Meanwhile *Quetzalcoatlus*, the biggest ever pterosaur, soared overhead. Mosasaurs and long-necked plesiosaurs were the great sea-going reptiles of their age, and *Pteranodon* swooped to pluck fishes from the waves.

About 65 million years ago, dinosaurs, pterosaurs, big marine reptiles, ammonites, and many other groups became extinct. Experts still dispute the cause of this catastrophe that brought the Mesozoic Era to a close.

Palaeocene Epoch

The Palaeocene ("old recent life") of 65–54 million years ago marks the first epoch of the Tertiary Period occupying most of the Cenozoic Era ("age of recent life"). Retreating seas exposed dry land in much of inland North America, Africa, and Australia. But South America was cut adrift with its own unique evolving "ark" of mammals. Everywhere, new kinds of mammal were appearing. Primitive early species waned as more advanced placentals took their place: condylarths (the first hoofed herbivores), rodents, and squirrel-like primates shared their world with bulky amblypods and primitive, early flesh-eating creodonts. Carnivorous mammals met some competition from big flightless birds of prey like *Diatryma*. Most fossil mammals come from North America, Europe, and Central Asia; in other places no extensive land-based sediments were laid down at this time.

At sea, gastropods and bivalves replaced ammonites as the leading molluscs. New kinds of sea urchin and foraminiferan replaced old ones. Among fishes, sharks seem to have been particularly plentiful.

Palaeocene world
This world map shows land masses in Palaeocene time. Lines represent Equator, Tropics, and Polar Circles.

Palaeocene life
These fossil animals represent some of those that flourished in Palaeocene time. They are not shown to scale.

INVERTEBRATES
a   Tentaculate

AMPHIBIANS
b   Caecilian

MAMMALS
c   Amblypod
d   Dermopteran
e   Primate
f   Condylarth
g   Multituberculate
h   Perissodactyl
i   Rodent

Eocene Epoch

Mountains rose and fissures leaked great lava flows in India and Scotland in the Eocene or "dawn of recent life" (54–38 million years ago). The rifting North Atlantic cut off North America from Europe, and South America lost links with Antarctica. Seas invaded much of Africa, Australia, and Siberia. Climates were generally warm or mild. Tropical palms even flourished in the London Basin.

Mammals continued to diversify. The first whales and sea cows swam in seas. Rodents ousted multituberculates as the main small mammals. Insectivores gave rise to bats. Primates included forest-dwelling ancestors of today's lemurs and tarsiers. Ungainly uintatheres stomped around North America and Asia. But condylarths were giving way to more modern ungulates – early horses, tapirs, and rhinoceroses, and pig-like anthracotheres in Asia and Europe. Ancestors of elephants roamed Africa. Meanwhile, the isolated condylarths of South America produced a unique zoo of hoofed mammals, along with edentates and marsupials. Australia's mammal fauna at this time remains a mystery.

Eocene world
This world map shows land masses in Eocene time. Lines represent Equator, Tropics, and Polar Circles.

Eocene life
These fossil animals represent some of those that flourished in Eocene time. They are not shown to scale.

INVERTEBRATES
a Coelenterate
b Nematode worm

MAMMALS
c Bat
d Tillodont
e Primate
f Creodont
g Carnivore
h Condylarth
i Amblypod
j Sea cow
k Proboscidean
l Perissodactyl
m Artiodactyls
n Edentate
o Whale

BIRDS
p Ratite
q Shore bird

©DIAGRAM

221

Oligocene Epoch

The Oligocene or "few recent" (kinds of life) lasted from 36–26 million years ago. Australasia had hived off from Antarctica, and left it isolated by ocean. This cooled world climates everywhere. Grasses and temperate trees ousted tropical vegetation from large areas. Grazing and browsing mammals multiplied – beasts like horses, camels, and rhinoceroses; while brontotheres ranged over Asia and North America. Dogs, stoats, cats, pigs, and rat-like rodents were on the increase. Africa was home to mastodonts, creodonts, hyraxes, anthracotheres, and the ape-ancestor *Aegyptopithecus*. Meanwhile isolated South America produced sloths, armadillos, rodents resembling guinea pigs, elephant-like pyrotheres, and others. While these creatures flourished, old fashioned hoofed and flesh-eating mammals – the condylarths and creodonts – were on the wane. Meanwhile at sea, early whales died out, largely replaced by toothed whales.

Oligocene world
This world map shows land masses in Oligocene time. Lines represent Equator, Tropics, and Polar Circles.

Oligocene life
These fossil animals represent some of those that flourished in Oligocene time. They are not shown to scale.

INVERTEBRATES
a Crustacean

MAMMALS
b Primates
c Creodont
d Embrithopod
e Pyrothere
f Perissodactyls
g Artiodactyls

BIRDS
h Coraciiform
i Apodiform

Miocene Epoch

The Miocene or "less recent" (with fewer modern creatures than the next epoch) lasted from 26–7 million years ago, longer than any other epoch. The world changed greatly now: ice covered Antarctica; the Mediterranean Sea dried up; India crashed into Asia; the Himalayas, Rockies, and Andes rose. But sea still isolated South America and Australia. Grasslands spread extensively and mammals reached their richest-ever variety. Many were hoofed grazers or browsers. Thus North America had horses, oreodonts, rhinoceroses, pronghorns, camels, protoceratids, and chalicotheres, with bear-dogs and sabre-toothed cats among the predators. Eurasia's "zoo" included early deer and giraffes, while African mammals included mastodonts, apes, and Old World monkeys. Great migrations saw elephants spread out from Africa to Eurasia and North America. Cats, giraffes, pigs, and cattle went the other way – from Eurasia to Africa. Horses found their way from North America into Eurasia. Meanwhile glyptodonts, armadillos, anteaters, New World monkeys, and horse-like litopterns evolved in isolated South America. Australia's Miocene marsupials and monotremes are little known.

Miocene world
This world map shows land masses in Miocene time. Lines represent Equator, Tropics, and Polar Circles.

Miocene life
These fossil organisms represent some of those that flourished in Miocene time. They are not shown to scale.

PLANTS
a Grass

MAMMALS
b Primates
c Carnivores
d Proboscidean
e Litoptern
f Notoungulate
g Perissodactyls
h Artiodactyls
i Monotreme

BIRDS
j Pelecaniform
k Ratite

©DIAGRAM

223

Pliocene Epoch

The Pliocene ("more recent") of 7–2 million years ago ended the first, long, Tertiary Period of the Cenozoic Era. Continents had taken up their present-day positions, and land linked North and South America. Antarctica's ice cap and new ones in the northern hemisphere cooled lands and oceans. Vegetation was like today's. Grasslands replaced many forests, so grazing mammals spread at the expense of browsers. Cattle, sheep, antelopes, gazelles, and other bovids reached their peak in Old World lands. North American mammals included horses, camels, deer, pronghorns, peccaries, mastodonts, beavers, weasels, dogs, and sabre-toothed cats. Rhinoceroses and protoceratids died out in North America. But ground sloths and other mammals moved in from South America. Meanwhile, dogs, bears, horses, mastodonts, and others colonized South America from the north. Early elephants, antelopes, and the ancestors of man roamed Africa. But isolated Australia's only newcomers were rodents, rafting in on mats of vegetation drifting south from Indonesia.

Pliocene time
This world map shows land masses in Pliocene time. Lines represent Equator, Tropics, and Polar Circles.

Pliocene life
These fossil animals represent some of those that flourished in Pliocene time. A re-dating of this period may put some items earlier. They are not shown to scale.

MAMMALS
a Marsupial
b Primates
c Desmostylan
d Proboscidean
e Notoungulate
f Litoptern
g Perissodactyl
h Artiodactyls
i Edentate
j Rodent
k Lagomorph

BIRDS
l Falconiform

Pleistocene Epoch

The Pleistocene ("most recent") Epoch of the short Quaternary Period lasted from about 2 million to 10,000 years ago. Ice Age cold gripped northern lands as ice caps and glaciers waxed and waned. Advancing cold forced creatures south, though some returned in intervals of warmth. So much water lay locked up in ice that the level of the oceans fell. Horses, camels, deer, tapirs, mastodonts, mammoths, dogs, and sabre-toothed cats lived in North America (though horses and camels died out there). Beasts migrating into South America helped wipe out many of its native creatures. Meanwhile, monkeys, hyaenas, hippopotamuses, and straight-tusked elephants thrived in warm-phase Europe. Eurasia's cold-adapted beasts included woolly mammoth, woolly rhinoceros, cave bear, and cave lion – all now extinct. Australia was home to outsize marsupials including *Diprotodon*, a giant kangaroo, and a marsupial "lion". Spreading probably from Africa, mankind evolved efficient hunting skills. Maybe this explains the disappearance of most large mammals before the Pleistocene ended. About 10,000 years ago Ice Age cold gave way to the long warm phase we call the Holocene or Recent Epoch – the time we live in now.

Pleistocene life
These fossil animals represent some of those that flourished in Pleistocene time. They are not shown to scale.

MAMMALS
a Marsupials
b Primates
c Carnivores
d Proboscidean
e Perissodactyls
f Artiodactyls
g Edentate
h Rodent

BIRDS
i Ratite

©DIAGRAM

Chapter 10

FOSSIL HUNTING

What we know about past life we owe to the patient work of teams of experts – between them skilled at finding, retrieving, preserving, displaying, and explaining fossils. This last chapter shows would-be fossil-hunters briefly what most of these techniques involve (and incidentally how some help to bring us such benefits as coal and oil).

We end with a short survey of pioneer palaeontologists and a worldwide museum guide to fossil collections. Not all of these are open regularly to the public, so check access before a visit.

Top-hatted palaeontologists direct fossil excavations in this lithograph from *Geology of Sussex* (1827), by the pioneer British dinosaur hunter Gideon Mantell. From Tilgate Forest's Early Cretaceous sandstone rocks Mantell obtained the bones of fossil dinosaurs, crocodiles and plesiosaurs. (British Museum – Natural History.)

Finding fossils

Fossil hunters search where man or weather has exposed sedimentary rocks – particularly limestones, shales, and clays; sometimes sandstone too. Likely sites are sea cliffs, quarries, disused mines road cuttings, building excavations, spoil heaps, streamsides with exposed bedrock, deserts, polar wastes, and mountainsides. Experienced collectors go armed with information obtained from local museums, guide books, and geological maps. They obtain landowners' consent; collect only in permitted areas (avoiding sites protected for their fossil rarities); and beware cliff falls, particularly after rain. Protective helmets may be needed, as well as old clothes and sturdy shoes, or rubber boots if working in soft mud.

Fossil hunters pace slowly, scanning gullies, cliffs, or piles of weathered rock. Some rocks bristle with fossils; others seemingly hold none. But patient

Rocks of the ages
This map shows where in the British Isles fossil hunters find fossil-bearing sedimentary rocks of different ages. Few fossils occur in igneous rocks such as basalt and granite, which are formed from upwelling molten rock, or in metamorphic rocks such as slate and marble, which are changed by tremendous heat or pressure.

■ Upper Cenozoic rocks

■ Lower Cenozoic rocks

■ Mesozoic rocks

■ Upper Palaeozoic rocks

■ Lower Palaeozoic rocks

□ Igneous and metamorphic rocks

searching and a practised eye could well reveal
tell-tale shiny or discoloured shapes in rock. Whole
fossils are a rarity. Collectors mostly find just
scattered teeth or broken bits of fossil leaf or bone.
Yet even fragments can betray the organisms they
belonged to. For example, short columns made of
disc-like plates could be the broken stems of
crinoids, or "sea lilies". The broken plates of all
echinoderms glint as they catch the light. Distinctive
teeth help experts to identify fossil sharks and
mammals. Deeply pitted bony plates come from
crocodiles; less deeply pitted plates, from turtles.
Turtle plates are thicker than those from certain fossil
fishes. Fossil fishes survive mostly just as scales and
individual bones. Our next two pages show how
fossils small and large can be extracted from the
rocks.

Fossil finds
Fossil hunters typically find
specimens like those shown here.
All but one are from marine
invertebrates, the most
widespread fossils, and most are
broken, worn, or incomplete.
a Upper Cenozoic brachiopod
b Lower Cenozoic shark's tooth
c Mesozoic ammonite
d Upper Palaeozoic crinoid
e Lower Palaeozoic graptolite

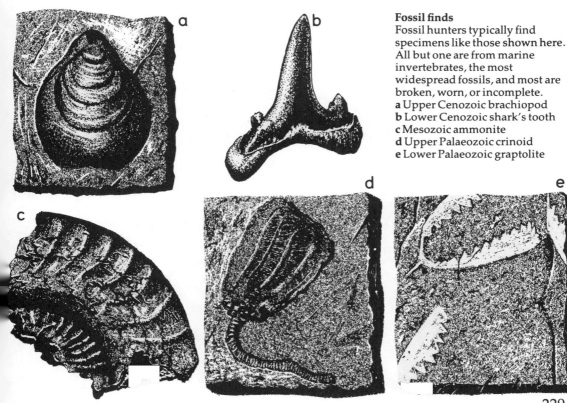

©DIAGRAM

229

Extracting fossils

Finding fossils is just the first stage of collecting. The fossil hunter must free fossils from their rocks, make a record of his finds, and transport these safely home.

You can go fossil collecting with just a geological hammer, hand lens, and old newspapers for wrapping finds. But collectors often take much more, inside a rucksack with extra space for specimens. A broad-bladed chisel called a bolster will split rocks along their bedding planes, revealing hidden fossils. Cold chisels help to cut them free. A flat-bladed trowel is helpful in soft rock. Old kitchen knives, brushes, even picks and spades, can have their uses. Sieves help you separate small teeth and bones from even smaller particles of clay or sand. The hand lens helps you identify tiny but important features. A steel rule serves for making measurements.

To remove a fossil from a rock, trim away as much rock as possible. Leave the rest for careful work at home. If your fossil breaks, mark joins and number

Fossil finder's toolkit
Dedicated fossil hunters may use all these items.

1 Rucksack
2 Geological map
3 Notebook
4 Marking pen
5 Compass
6 Hand lens
7 Geological hammer
8 Punch
9 Cold chisel
10 Bolster
11 Trowel
12 Kitchen knife
13 Toothbrush
14 Rule
15 Newspapers
16 Sticky tape
17 Paper tissues
18 Plastic bags
19 Boxes

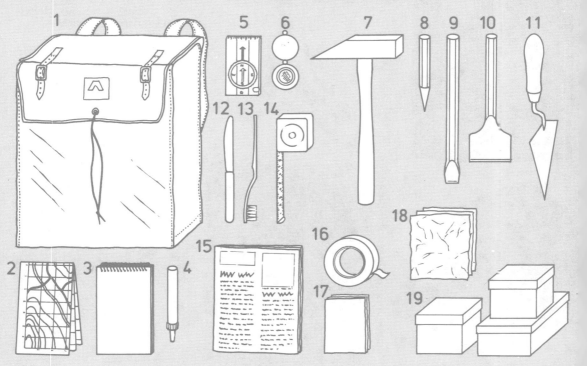

the pieces, using waterproof ink or a felt-tip pen. Wrap together fragments for later reassembly. Use a water- or solvent-based hardening solution to strengthen fragile fossils in soft clay or sand. When specimens have dried, wrap them in paper tissues, kitchen foil, moss, or sand, and place them in tins or boxes. Stronger specimens can go in polythene or linen bags.

Gather samples of loose rock and sieve them back at home for tiny fossil teeth, bones, or seeds. Number each item, then use a notebook to record its number, name, locality, and details of its parent rock. A sketch or photograph of rock layers at the site gives useful future reference.

Collecting big fossils, such as bones of dinosaurs, calls for special skills and teams of workers. Report such finds to the palaeontology department of a museum or university.

Fossil hunting
We show fossil hunters working near a cliff foot.
A This fossil hunter is finding fossils in lumps of rock. He cracks open ball-shaped lumps, and splits shale along its bedding plane. A helmet protects his head from falling stones.
B Another fossil hunter sketches and labels the nature, depth, and fossil content of exposed rock layers and their fossil beds.

1 Brown limestone
2 Fossil-oyster bed
3 Brown limestone
4 Hard sandstone: no fossils
5 Green shale: containing fossil nautiloids (**a**) and fossil trilobites (**b**)
6 Dark shale
7 Green shale
8 Talus: fallen rocks concealing lower rock beds

Cleaning and repairing fossils

Museums receive many fossil skeletons resembling unsorted pieces of a jig-saw puzzle stuck in rock. But there are ways to free each bone or bit of shell and clean it for display.

The first task is soaking, sawing, slicing, or otherwise removing protective packing wrapped around a fossil still embedded in its rock. Next, technicians may use special chemical solutions to harden exposed, fragile bones. Then they set to work with tools or chemicals, or both. People may chip away with hammer and chisel. But experts can work faster and more carefully with help from power tools, such as grinding burrs, or dental drills with rapidly revolving diamond cutting wheels. Vibrating tungsten points of pneumatic power pens, and gas jets firing an abrasive powder can cut through rock as if it were as soft as butter. Ultrasonic waves attack weaknesses in certain rocks. But sewing needles serve for cleaning tiny, fragile skulls.

Sometimes laboratory workers crush rock with a pestle and mortar, sieve the particles, and then inspect them with a microscope. All this separates and shows up microfossils.

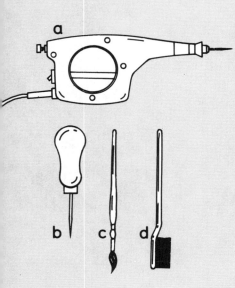

Cleaning tools (above)
Of tools shown, the first two free fossils from hard rock; the others remove soft rock from fossils.
a Speed engraver, with a fast-vibrating point
b Awl
c Fine-bristle brush
d Toothbrush

Acid preparation (right)
Illustrations show four stages in using acid to reveal a fossil embedded in a rock.
1 Rock encloses almost all the fossil.
2 The rock is soaked in acid for 2- to 6-hour periods.
3 Each time the soaked rock has been removed from the acid, it is washed in deionized water for a day, completely dried, and then the exposed fossil is painted with a plastic glue.
4 The prepared specimen is left to soak in water for up to two weeks.

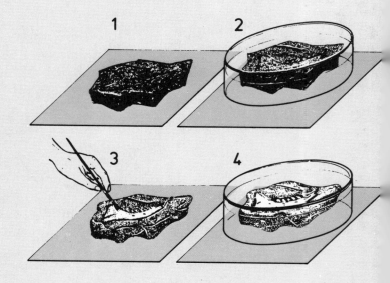

Chemicals have uses, too. Soaking limestone in dilute acetic or formic acid may remove the rock without dissolving fossils in it. Other acids attack rocks rich in iron and silica, while certain alkalis will break down shales and clays. Some chemicals are poisonous or burn the skin, so people handle these with special care.

Mending broken fossil bones and teeth is another job for the laboratory. Workers match two pieces at a time. They clean matching surfaces, then stick them together with a special adhesive. Rubber bands, metal clamps, or other aids hold the bits together until the adhesive dries.

Fossils rich in iron pyrites call for more than cleaning or repair. They need protection from decay. Damp air rots pyritic fossils, so museums store them in dry air. Treating already damaged specimens with ammonia arrests decay; and washing in fresh water prevents crumbling of pyritic fossils found at seashore sites. Even so, "pyrite disease" remains a major problem.

Inside a fish's skull
This drawing of an ancient fish's brain is based on one made by the Swedish palaeontologist Erik Stensiö in the 1920s. Stensiö used fine needles to remove the fossil skull, revealing rock formed in internal cavities that once contained brain, nerves, and blood vessels. Stensiö's discovery proved cephalaspids had been early fishes, not salamanders as some people once supposed.

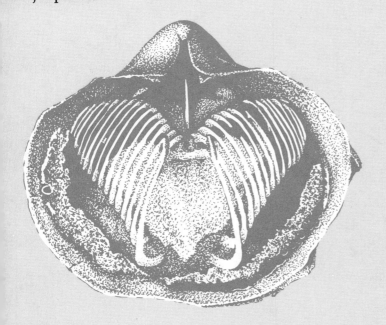

Inside a brachiopod
This much enlarged view shows the inside of one valve of a fossil brachiopod, after acid treatment. Acid has etched away unwanted rock, revealing spiral "rib-like" structures that once supported the creature's feeding system.

Fossils for display

Sometimes, museum workers extract most of a fossil skeleton from rock. The museum might then decide to reconstruct the skeleton and mount it for display. This calls for expert knowledge of how bones or bits of shell fitted together in the creature when it was alive. Anatomists arrange bones in order, and judge their angles. Intelligent guesswork is needed for some bones of unfamiliar beasts. Then technicians can rebuild the skeleton in a life-like pose. They may set up the hip bones first, then most vertebrae, then skull, ribs, breastbone, limb bones, and the tail.

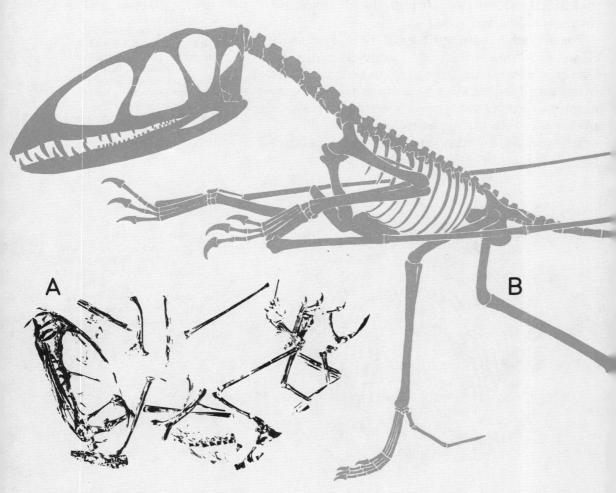

A

B

234

Different techniques serve for mounting specimens of different types and sizes. Workers hang heavy dinosaur bones by ropes from wooden scaffolding. They link the bones by means of angled metal pipes or rods, fastened to the bones with metal clips. Then they remove the scaffolding, leaving the skeleton exposed to view. This task can take a team of workers several months.

Museums also produce life-size, lightweight copies of important fossil skeletons. They make a plaster mould of every bone, then fill each mould with fibreglass. Museums exchange such casts to increase the variety of specimens they show.

From mounted skeletons, artists can draw or model fossil creatures as they would have looked in life. Important aids are the tell-tale bumps, grooves, and scars on bones, showing where muscles were attached. Experts studying such clues can work out shapes, weights, and volumes of beasts that lived hundreds of millions of years ago.

Dimorphodon restored
Illustrations picture three stages of a "resurrection" of a Jurassic pterosaur, *Dimorphodon macronyx*.
A *Dimorphodon*'s fossil bones as found scattered in Lower Lias rock near the southern English seaside town of Lyme Regis.
B Reconstruction of the whole fossil skeleton, about 1m (3ft 3in) long. Knowledge of anatomy enabled experts to infer shapes, sizes, and positions of any missing bones.
C Restoration of *Dimorphodon* in life-like pose.

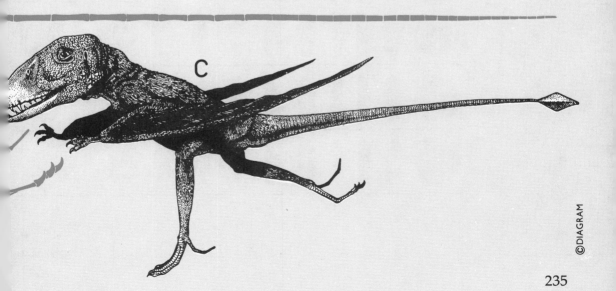

C

©DIAGRAM

Using fossils

Collecting and displaying fossils gives amateurs a fascinating glimpse of prehistoric life. For professional geologists, fossils mean much more. They offer clues to past climates, extinct communities, and the extent and depth of different rock formations. Indirectly, certain fossils even serve as guides to sedimentary rocks rich in useful minerals or fuels.

Macrofossils (big fossils) such as brachiopods, graptolites, and trilobites are among the chief guides to Palaeozic sedimentary rocks. Molluscs are their Mesozoic and Cenozoic counterparts. But for these latter eras, foraminiferans, ostracods, and other microfossils matter more.

Teams of geologists recover microfossils from sediments obtained as deep-drilled cores. In the laboratory, experts sieve and concentrate the samples. Then they identify these with a powerful microscope and "field guides" to microfossils. It is

Correlating rocks
Fossil foraminiferans showing rapid evolutionary change helped palaeontologists correlate these Jurassic rock strata in Montana. In three strata nine fossil forms stayed the same, but others vanished or appeared. Thus each layer has some fossil feature enabling experts to identify that stratum wherever it occurs.
1 Siltstone stratum
2 Limestone and shale
3 Shale
4 Limestone and shale
5 Not exposed
6 Shale
7 Limestone

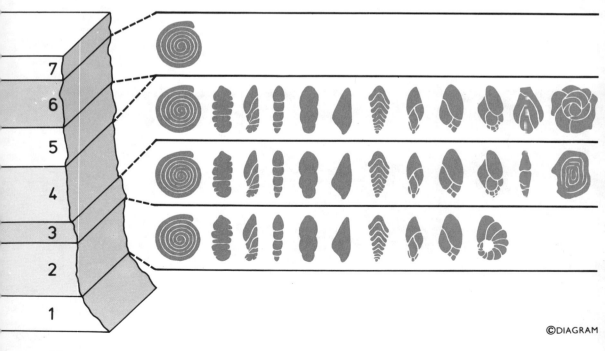

©DIAGRAM

possible to tell if a core came from a locality and depth likely to hold minerals or fossil fuels. All fossil fuels and certain other useful substances result from long-dead organisms, changed and concentrated in some way. For instance, coal forms slowly as a mass of plant material is fossilized, then buried under sediments and changed by heat and pressure. Petroleum and natural gas form when impermeable sea-bed sediments trap permeable rock containing the remains of billions of marine micro-organisms whose bodies once secreted oily droplets.

Limestone used for building blocks or in cement comprises fossil skeletons of algae and invertebrates that lived in shallow prehistoric seas.

Even some phosphate fertilizer comes from sea-bed layers enriched by phosphorus from fish bones and the shells of sea-bed creatures.

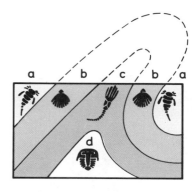

Fossils and folding (above)
Fossils of beasts that lived at different times help experts to date rocks that have been tilted or even overturned by folding.
a Devonian eurypterids
b Silurian brachiopods
c Ordovician crinoid
d Cambrian trilobite

Fossil fuel formation (left)
Block diagrams show parts played by prehistoric life forms in coal and oil formation.
1 Coal formation
a Rotting swamp plants form peat.
b Pressure of accumulating sediments turns peat to lignite.
c Increased pressure turns lignite to bituminous coal.
d Heat and deformation change bituminous coal to anthracite.
2 Oil formation
a Marine micro-organisms die and fall to the sea bed.
b Pressure cooking and bacteria act on micro-organisms to release hydrocarbons.
c Hydrocarbons migrate up through porous rock until trapped by impervious rocks here tilted by a risen salt dome.

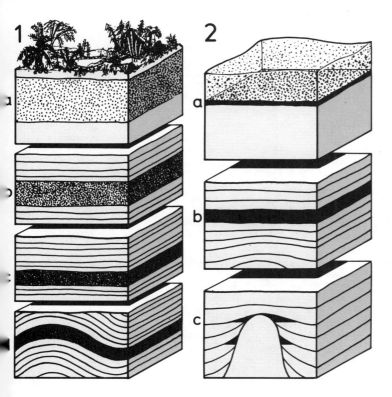

Famous fossil hunters 1

These four pages list major fossil hunters and others whose achievements added significantly to our understanding of prehistoric animals and plants. No list this brief could be complete; ours stresses pioneers, especially from the Western World. Outstanding personalities have also come from the USSR, Poland, China, India, and elsewhere.

Roy Chapman Andrews

Mary Anning

William Buckland

Agassiz, Jean Louis (1807–73) Swiss-born naturalist who made key studies of fossil fishes and showed that ice caps had covered much of Pleistocene Europe and North America.

Alberti, Friedrich August von (1795–1878) German geologist who in 1824 named the Triassic system from a threefold division of rocks found in Germany.

Ameghino, Florentino (1854–1911) Argentinian palaeontologist who described many South American fossil dinosaurs and mammals, largely from specimens collected by his brother Carlos.

Andrews, Roy Chapman (1884–1960) American leader of expeditions for the American Museum of Natural History. His Gobi Desert journeys of the 1920s produced the first-known dinosaur eggs.

Anning, Mary (1799–1847) British fossil collector from Lyme Regis, Dorset. She found the first British pterosaur and the first complete ichthyosaur and plesiosaur.

Bakker, Robert American palaeontologist who in the 1970s argued, controversially, that dinosaurs had been warm-blooded and that birds are dinosaurs.

Barrande, Joachim (1799–1883) French pioneer researcher of Palaeozoic rocks and fossils. His 32-volume work included the first accounts of 4000 fossil species.

Beyrich, Heinrich Ernst (1815–96) German palaeontologist who coined the term Oligocene in 1854.

Brongniart, Adolphe (1801–76) French botanist who founded palaeobotany. In 1822 he published the first account of all known fossil plants.

Brongniart, Alexandre (1770–1847) French geologist who in 1829 named the Jurassic Period from its limestone rocks in the Jura Mountains. With Georges Cuvier he compiled a chronological sequence of Tertiary rocks in France.

Bronn, Heinrich Georg (1800–62) German palaeontologist and geologist who laid the basis for a chronological study of fossil organisms in Germany.

Broom, Robert (1866–1951) Scottish palaeontologist who made major discoveries about mammal-like reptiles and the origin of mammals.

Brown, Barnum (active 1890s and early 1900s) American fossil hunter who pioneered the Canadian "dinosaur rush", collecting bones from the Red Deer River Valley in 1910.

Buckland, William (1784–1856) British geologist who in 1824 described *Megalosaurus*, the first dinosaur to get a scientific name.

Buffon, Georges, Comte de (1707–88) French naturalist. He helped pioneer the idea that a succession of plants and animals dated back farther than theologians believed.

Hawkins' Iguanodon
This is a model of Benjamin Waterhouse Hawkins' restoration of *Iguanodon*. The full-size cement, stone and metal original still stands in a London park.

Conybeare, William Daniel (1787–1857) British geologist who in 1822 named the Carboniferous system. He also first described *Ichthyosaurus*.

Cope, Edward Drinker (1840–97) American zoologist, who described many American fossil fishes, reptiles, and mammals – especially from the newly opened West.

Cushman, Joseph (1881–1949) American palaeontologist who pioneered the use of foraminiferans as guides to the relative ages of certain rocks.

Cuvier, Georges (1769–1832) French anatomist and palaeontologist who pioneered the scientific study of fossil vertebrates. He believed groups of prehistoric beasts had perished from a series of natural catastrophes.

Dart, Raymond Arthur (1893–) Australian anatomist who in 1925 first described an *Australopithecus* fossil, from Botswana.

Darwin, Charles Robert (1809–82) British naturalist who proposed the theory of evolution by natural selection.

Deshayes, Gérard Paul (1797–1875) French conchologist, whose studies of fossil shells laid a basis for subdividing the Tertiary Period into epochs.

Desnoyers, Jules French geologist who in 1829 separated the Quaternary Period from the Tertiary Period. (Tertiary was a term coined in the mid 18th century.)

Douglass, Earl (1862–1931) American fossil collector. In 1909 he found a mass of embedded dinosaur bones in Utah (see p. 246 – Jensen, Utah: Dinosaur National Monument).

Dubois, Eugène (1858–1940) Dutch anatomist and palaeontologist who in 1891 made the first find of a *Homo erectus* fossil.

Ehrenberg, Christian (1795–1876) German naturalist who pioneered the study of microfossils. His *Mikrogeologie* (1854) depicted radiolarian skeletons and showed the role of microfossils in building limestones such as chalk.

Fischer von Waldheim, Gotthelf (1771–1853) German scientist who helped pioneer palaeontology in Russia.

Gilmore, Charles Whitney (1874–1945) American museum curator and expedition leader. He greatly enlarged the fossil reptile collection of the American National Museum of Natural History.

Hall, James (1811–98) Main founder of the "American school" of palaeontology. He wrote on the Palaeozoic invertebrates of New York State.

Halloy, Jean-Baptiste-Julien Omalius d' (1783–1875) French geologist who in 1822 named the Cretaceous Period.

Hawkins, Benjamin Waterhouse (1807–89) British sculptor who created the first life-size restorations of dinosaurs, completed in 1854.

Edward Drinker Cope

Georges Cuvier

Charles Darwin

Famous fossil hunters 2

Huene's Plateosaurus
Plateosaurus, probably a
quadruped, appears bipedal in
many restorations. Friedrich von
Huene's discoveries added to
our knowledge of this
prosauropod dinosaur.

James Jensen with the shoulder
blade of "Supersaurus"

Othniel Charles Marsh

Hooke, Robert (1635–1703) English scientist who gave the first descriptions of fossil wood, and foreshadowed the idea of fossils as clues to evolution.

Huene, Friedrich von (active early 1900s) German palaeontologist. In 1921 he found a fossil herd of *Plateosaurus* dinosaurs near Trossingen in Germany.

Hutton, James (1726–97) Scottish geologist who showed that natural agents still at work had produced geological changes over vast periods of time.

Hyatt, Alpheus (1838–1902) American geologist and palaeontologist who helped to classify ammonites.

Janensch, Werner (active early 1900s) German palaeontologist whose expeditions of 1909–12 recovered a wealth of dinosaurs from Tendaguru, now in Tanzania.

Jensen, James American fossil hunter who found two of the largest known dinosaurs ("Supersaurus" and "Ultrasaurus") in Colorado in the 1970s.

Lamarck, Jean-Baptiste de (1744–1829) French palaeontologist, pioneer in the scientific study of fossil invertebrates.

Lapworth, Charles (1842–1920) British geologist who, using graptolites, identified the Ordovician system in 1873.

Leakey, Louis Seymour (1903–72) British palaeontologist who worked in East Africa, with his wife Mary and son Richard. Their finds have added much to our knowledge of man's early evolution.

Leidy, Joseph (1823–91) American anatomist who pioneered the study of fossil vertebrates in North America. In 1856 he became the first to name an American dinosaur.

Lhuyd, Edward (1660–1709) Welsh natural historian who in 1699 produced the first book about British fossils.

Linnaeus, Carolus (1707–78) Swedish botanist who established a basis for classifying living things.

Lyell, Sir Charles (1797–1875) British geologist whose *Principles of Geology* helped found the modern science of geology. In the 1830s he named the Eocene, Miocene, Pliocene, and Pleistocene epochs.

Mantell, Gideon Algernon (1790–1852) British amateur geologist who in 1825 described *Iguanodon*, the second dinosaur to be named.

Marsh, Othniel Charles (1831–99) American palaeontologist whose expeditions discovered scores of fossil vertebrates in the West and Mid West. He described countless fossils including 17 dinosaur genera that still bear the names he gave them.

Murchison, Sir Roderick Impey (1792–1871) British geologist who established the sequence of early Palaeozoic rocks. He identified the Silurian and Permian systems.

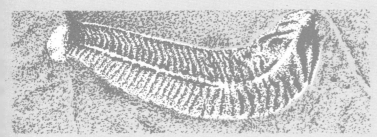

Sprigg's Spriggina
The name of the ancient sea
worm *Spriggina* commemorates
Sprigg's remarkable discoveries
of Precambrian fossil animals in
South Australia.

Nopcsa, Baron Franz (1877–1933) Palaeontologist born in what is now Romania; he discovered or described a number of dinosaurs.

Orbigny, Alcide Charles d' (1802–57) French naturalist whose study of fossil invertebrates revealed regional distributions of species in ancient seas.

Osborn, Henry Fairfield (1857–1935) American palaeontologist and expedition leader who organized major fossil hunts in Colorado, Wyoming, and Central Asia. He wrote over 600 scientific papers, and became president of the American Museum of Natural History.

Owen, Sir Richard (1804–92) British anatomist and palaeontologist who coined the name "Dinosauria". He wrote the first work on general palaeontology in English, and became the first director of the British Museum (Natural History).

Sir Richard Owen

Phillips, John (1800–74) British geologist who in 1840 named the Mesozoic and "Kainozoic" (Cenozoic) eras.

Romer, Alfred Sherwood (1894–1973) American palaeontologist who contributed new ideas about how fishes, amphibians, and reptiles evolved. He wrote major textbooks on fossil and living vertebrates.

Scheuchzer, Johann Jakob (1672–1733) Swiss botanist who wrote one of the first books to picture fossil plants.

Schimper, Wilhelm Philipp (1808–80) German palaeontologist who in 1874 named the Palaeocene Epoch.

Schlotheim, Ernst von (1764–1832) German palaeontologist who helped pioneer the use of fossils as clues to the relative ages of rock strata.

Sedgwick, Adam (1785–1873) British geologist who named the Palaeozoic Era, in 1838, and the Cambrian and Devonian systems.

Seward, Sir Albert Charles (1863–1941) British palaeobotanist who wrote key books on fossil plants.

Smith, William (1769–1839) British geologist known as the father of English geology. He identified layers of sedimentary rock by their fossils, and in 1815 made the first geological map of England and Wales.

Alfred Sherwood Romer

Sprigg, R.C. Australian geologist who in 1947 discovered major Precambrian fossils in South Australia.

Sternberg, Kasper von (1761–1838) Czech palaeobotanist who related the classification of fossil plants to that for living plants.

Walcott, Charles (1850–1927) American palaeontologist who made important studies of trilobites.

Woodward, Arthur Smith (1864–1944) British palaeontologist who made a major catalogue of fossil fishes, with many kinds reclassified.

Woodward, John (1665–1728) English palaeontologist who compiled one of the first classifications of fossils.

Zittel, Karl Alfred von (1830–1904) German scientist who compiled major handbooks of fossils, and a key history of palaeontology.

William Smith

©DIAGRAM

Museum displays 1

Thousands of museums show or store fossils or models of prehistoric animals and plants. In some Western countries the chief town in almost every state or county depicts local prehistoric life. These six pages give just a brief selection of interesting collections worldwide. Items appear in alphabetical order of country and then city.

A giant ammonoid in the South Australian Museum

A giant moa in the Natural History Museum, Vienna

ARGENTINA
Buenos Aires: Argentine Natural Science Museum Features a fine collection of fossil vertebrates from Argentina.
La Plata: Museum of La Plata University Includes fossil dinosaurs from Argentina.
AUSTRALIA
Adelaide: South Australian Museum Contains fossil mammals, reptiles (especially crocodiles), fishes, amphibians, etc.
Brisbane: Queensland Museum Includes fossil mammals, reptiles, fishes, and some fossil bird remains. Little on show before 1986.
Melbourne: Geological Museum Includes a big collection of fossil graptolites from Victoria.
Melbourne: Museum of Victoria Has mostly mammal fossils.
Perth: Western Australian Museum Includes fossil invertebrates and vertebrates (mostly mammals).
Sydney: Australian Museum Includes mostly Devonian fishes and mammals. Exhibition galleries of its Department of Palaeontology show the story of life on Earth.
AUSTRIA
Vienna: Natural History Museum Its 30,000 specimens include mounted skeletons of a moa, Austrian Ice Age mammals, and remains of an armoured dinosaur.
BELGIUM
Brussels: Royal Institute of Natural Sciences This has the world's best and biggest collection of *Iguanodon* skeletons. Other exhibits include mosasaurs and invertebrate Mesozoic and Cenozoic fossils.
CANADA
Calgary, Alberta: Zoological Gardens An outdoor park shows 50 full-size models of dinosaurs.
Drumheller, Alberta: Tyrrell Museum of Palaeontology On show are major fossils and restorations of Late Cretaceous dinosaurs from Alberta.
Edmonton, Alberta: Provincial Museum of Alberta Includes a partial dinosaur skeleton and life-size models of several other dinosaur genera.
Ottawa, Ontario: National Museum of Natural Sciences A "life through the ages" sequence includes mounted examples of western Canada's Late Cretaceous dinosaurs, and a reconstructed prehistoric forest.
Toronto, Ontario: Royal Ontario Museum Canada's largest public museum is rich in North American vertebrate fossils. The Dinosaur Gallery includes dinosaurs, mosasaurs, an ichthyosaur, a plesiosaur, and an early crocodile.

CHINA

Beijing (Peking): Beijing Natural History Museum On show are many Chinese fossils. The five dinosaur skeletons include an immense reconstructed *Shantungosaurus*. Museums in some other major Chinese cities also have significant fossil displays.

CZECHOSLOVAKIA

Prague: Národní Museum The national museum includes a number of fossils found in the 19th century. Many were the first of their kind to be discovered.

FRANCE

Paris: National Museum of Natural History This has the largest fossil collection in France. It includes bones or casts of dinosaurs from various continents. (Regional museums also feature fossil displays.)

GERMANY (EAST)

East Berlin: Natural History Museum, Humboldt University Its fossil collection has important dinosaur skeletons from Late Jurassic Tanzania; the *Brachiosaurus* is the world's largest mounted dinosaur.

GERMANY (WEST)

Darmstadt: Hesse State Museum Unusual exhibits include an American mastodont skeleton.

Frankfurt am Main: Senckenberg Natural History Museum On show is a wealth of vertebrate fossils, including ichthyosaurs, plesiosaurs, and dinosaurs from various parts of the world.

Munich: Bavarian State Institute for Palaeontology and Historical Geography This includes the first known skeleton of the Jurassic dinosaur *Compsognathus*.

Stuttgart: State Museum for Natural History Its collection includes important dinosaur fossils.

Tübingen: Institute and Museum for Geology and Palaeontology Fossils on show include dinosaur skeletons and casts.

GREECE

Athens: Department of Geology and Palaeontology, University of Athens This has important fossil Pliocene mammals.

INDIA

Calcutta: Geology Museum, Indian Statistical Institute The chief exhibit is the only mounted skeleton of the early sauropod *Barapasaurus*. (Other museums also feature fossils.)

ITALY

Bologna: G. Capellini Museum has a large dinosaur cast.

Genoa: Civic Museum of Natural History The collection includes fossil plants, invertebrates, and vertebrates.

Milan: Civic Museum of Natural History Fossils feature in its collection.

Padua: Museum of the Institute of Geology This museum includes fossil plants.

Rome: Museum of Palaeontology, Institute of Geology and Palaeontology This has fossil Quaternary mammals, footprints, and invertebrates.

JAPAN

Osaka: Museum of Natural History Fossil plants, invertebrates, and vertebrates contribute to a history in nature theme.

Tokyo: National Science Museum On show is a significant fossil collection, including a dinosaur display.

The mounted *Brachiosaurus* in an East Berlin museum

An ichthyosaur in Stuttgart's Natural History Museum

© DIAGRAM

An *Allosaurus* in Osaka's Museum of Natural History

243

Museum displays 2

KENYA
Nairobi: Kenya National Museum Includes remains of fossil man from East Africa.

MEXICO
Mexico City: Natural History Museum Includes a cast of the sauropod dinosaur *Diplodocus*.

MONGOLIA
Ulan-Bator: State Central Museum Features an imposing collection of dinosaurs from the Gobi Desert.

MOROCCO
Rabat: Museum of Earth Sciences This is planned to house fossils including the largest known skeleton of the sauropod dinosaur *Cetiosaurus*.

NIGER
Niamey: National Museum Fossils include *Ouranosaurus*, a big sail-backed iguanodontid dinosaur.

POLAND
Warsaw: Palaeobiology Institute, Academy of Sciences Has a major collection of Mongolian dinosaurs, but seldom on show.

SOUTH AFRICA
Cape Town: South African Museum This has important fossil reptiles from Permian and Triassic southern Africa.

SPAIN
Madrid: Natural Science Museum Includes a cast of the sauropod dinosaur *Diplodocus*.

SWEDEN
Uppsala: Palaeontological Museum, Uppsala University This has fossils of Chinese dinosaurs including a sauropod.

UNITED KINGDOM
Cambridge: Sedgwick Museum, Cambridge University Fossils include some local dinosaur remains.

Cardiff: National Museum of Wales Has a large fossil collection, including Palaeozoic Welsh invertebrates.

Cheddar, Somerset: Cheddar Caves Museum Features Pleistocene remains, including a Palaeolithic human burial.

Clitheroe, Lancs: Clitheroe Castle Museum Has a significant collection of Carboniferous fossils.

Dorchester: Dorset County Museum Includes dinosaur footprints.

Edinburgh: Royal Scottish Museum Has a major national collection of early fossils including invertebrates (notably crinoids and eurypterids), vertebrates, and plants.

Elgin: Elgin Museum Includes world-famous fossils of Scotland's Early Mesozoic reptiles.

Glasgow: Hunterian Museum Includes a *Triceratops* skull and paintings of Scottish Mesozoic reptiles.

Glasgow: Victoria Park Its Fossil Grove building houses remarkable fossil Carboniferous tree stumps.

Huddersfield: Tolson Memorial Museum Includes a section with geological exhibits.

Ipswich: Ipswich Museum Local invertebrate and vertebrate fossils, including bones, teeth, and tracks of dinosaurs.

Keighley, West Yorks: Cliffe Castle Has a geological gallery.

Leicester: Leicestershire Museum and Art Gallery Its fossils include remains of Precambrian invertebrates.

Ouranosaurus in the National Museum, Niamey

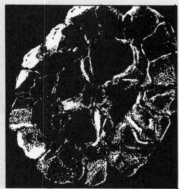

Coccolith in the Sedgwick Museum, Cambridge

A fossil tree stump in Glasgow's Victoria Park

London: British Museum (Natural History) One of the world's great natural history museums, with more than three million fossils on show or stored, plus many casts. Displays represent most major forms of prehistoric life.

London: Crystal Palace Park Has the world's oldest display of full-size model dinosaurs, completed in 1854.

London: Geological Museum Permanent exhibitions include "British Fossils" and "Britain Before Man".

Lyme Regis: Lyme Regis Museum *Ichthyosaurus* and other local Jurassic fossils figure in this small collection.

Maidstone: Maidstone Museum Includes some *Iguanodon* bones with a related display.

Manchester: Manchester Museum Fossils and minerals figure in its total collection of over eight million items.

Newcastle upon Tyne: Hancock Museum One of England's finest natural history museums. It includes geological items.

Oxford: University Museum The five thousand fossil exhibits include unique dinosaur specimens.

Peterborough: City of Peterborough Museum and Art Gallery Includes an exhibition of local geology.

Portsmouth: Cumberland House Natural Science Museum and Aquarium Has a full-size restoration of the dinosaur *Iguanodon*.

Sandown: Museum of Isle of Wight Geology Houses local fossils, including part of the oldest-known bone-headed dinosaur, *Yaverlandia*.

Sunderland: Sunderland Museum Has a sizable collection of Permian fossils, including fishes and the lizard-like gliding reptile *Weigeltisaurus*.

Tenby, Dyfed: Tenby Museum Has a significant geological collection on show.

Torquay: Natural History Society Museum On show are Prehistoric finds from Kent's Cavern.

York: Yorkshire Museum Extinct and fossil birds figure in this museum's collection.

UNITED STATES OF AMERICA

Amherst, Massachusetts: Amherst College Displays a major collection of dinosaur footprints.

Austin, Texas: Texas Memorial Museum On show are Late Palaeozoic and Mesozoic reptiles (including sea turtle, mosasaur, and dinosaurs), and Pleistocene mammals.

Berkeley, California: University of California Museum of Paleontology Includes Triassic and Jurassic reptiles.

Boulder, Colorado: University Natural History Museum On show are Jurassic dinosaur fossils.

Buffalo, New York: Buffalo Museum of Science Fossils include dinosaur bones, footprints, eggs, skin impressions, and other items.

Cambridge, Massachusetts: Museum of Comparative Zoology, Harvard University Displays a major collection of fossil vertebrates. The museum includes fossil fishes, the best North American collection of South America's Early Mesozoic amphibians and reptiles, and North American dinosaurs.

Canyon, Texas: Panhandle Plains Museum On show are local Triassic reptiles.

The London specimen of *Archaeopteryx*

A reconstruction of *Weigeltisaurus* from Sunderland Museum

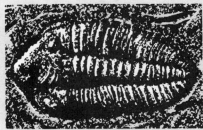

A trilobite in Harvard's Museum of Comparative Zoology

© DIAGRAM

Museum displays 3

Dunkleosteus's huge head on show in Cleveland, Ohio

A bird from the La Brea tar pits of Los Angeles

Chicago, Illinois: Field Museum of Natural History A major museum that includes fossil plants, invertebrates, and vertebrates from South America and the Western United States – for example mounted dinosaurs.

Cincinnati, Ohio: University of Cincinnati Fossils include some dinosaur remains.

Cleveland, Ohio: Natural History Museum This is famous for *Dunkleosteus* and other Devonian fossil fishes, the only mounted skeleton of the sauropod dinosaur *Haplocanthosaurus*, and Pleistocene mammals including a mastodont.

Denver, Colorado: Denver Museum of Natural History Its "Succession of Life" displays feature dinosaurs, marine reptiles, and 50 million years of mammal evolution.

Durham, North Carolina: North Carolina Museum The fossil collection includes some dinosaur bones.

East Lansing, Michigan: The Museum, Michigan State University A Hall of Life includes fossils and wall paintings from successive geological eras.

Flagstaff, Arizona: Museum of Northern Arizona Items include fossils of a small, early, armoured ornithopod dinosaur.

Fort Worth, Texas: Fort Worth Museum of Science and History Exhibits include mounted dinosaurs.

Hays, Kansas: Sternberg Memorial Museum Toothed birds are among its display of local Cretaceous fossils.

Houston, Texas: Houston Museum of Natural Science Large exhibits include much of a *Diplodocus* skeleton.

Jensen, Utah: Dinosaur National Monument Comprises more than 200,000 acres of fossil-rich canyons. The Carnegie Quarry is now a covered Visitor Center where you can view technicians freeing thousands of dinosaur bones from rock.

Laramie, Wyoming: W.H. Reed Museum Exhibits include part of a fossil sauropod, *Apatosaurus*.

Lawrence, Kansas: University of Kansas Museum of Natural History The collection includes Mesozoic fossils.

Lincoln, Nebraska: University of Nebraska State Museum This has a good display of fossil mammals.

Los Angeles, California: Los Angeles County Museum of Natural History Holds a big collection of Cretaceous fossil vertebrates (e.g. *Pteranodon, Tylosaurus*) and the world's largest collection of later Pleistocene fossil vertebrates. At its La Brea Center viewers can see experts excavating prehistoric bones from tar pits.

Newark, Delaware: University of Delaware Fossils include some sauropod dinosaur remains.

New Haven, Connecticut: Peabody Museum of Natural History, Yale University This has a major collection of fossil vertebrates – especially American dinosaurs and early mammals.

New York City, New York: American Museum of Natural History This great collection has a rich display of mounted skeletons – including fishes, amphibians, reptiles, and mammals. No other museum contains so many dinosaurs. Exhibits include fossil eggs, tracks, and skin imprints.

Norman, Oklahoma: Stovall Museum, University of Oklahoma Fossils include a big, flesh-eating dinosaur from Oklahoma.

Palaeoscincus in the American Museum of Natural History

A fossil forest
Logs literally turned to stone lie jumbled in Rainbow Forest. This is one of the six fossil forest areas in Arizona's Painted Desert.

Painted Desert Arizona: Petrified Forest National Park Has the world's largest concentration of petrified wood – six "forests" of 150-million-year-old giant conifer logs, transformed to agate and chalcedony.

Peoria, Illinois: Lakeside Museum and Art Center Fossils include sauropod dinosaur remains.

Philadelphia, Pennsylvania: Academy of Natural Sciences This has some of the first fossil dinosaurs found in North America.

Pittsburgh, Pennsylvania: Carnegie Museum of Natural History On show are major fossil displays, including a Mesozoic Hall with some of the world's best-preserved, mounted specimens of Late Jurassic dinosaurs, as well as mosasaurs and a marine turtle.

Princeton, New Jersey: Museum of Natural History, Princeton University Fossils include Late Cretaceous dinosaurs and one of the first-known fossil bats.

St Paul, Minnesota: The Science Museum of Minnesota Contains local Cretaceous fossils, including champsosaurs.

Salt Lake City, Utah: Utah Museum of Natural History Fossils include dinosaurs from the famous Cleveland-Lloyd Quarry, also pterosaur tracks.

San Francisco, California: California Academy of Science Fossil items include single dinosaur bones.

Scranton, Pennsylvania: Everart Museum Its fossils include some dinosaur bones.

Vernal, Utah: Utah Natural History State Museum Exhibits include a *Diplodocus* skeleton.

Washington, D.C.: National Museum of Natural History, Smithsonian Institution This major museum has fossils representing most phases of prehistoric life. The Department of Paleobiology holds one of the world's largest collections of type specimens (the first fossils of their kind to get a description and name).

USSR

Leningrad: Central Geological and Prospecting Museum Fossil exhibits include an Asian hadrosaurid dinosaur.

Leningrad: Museum of Zoology Unique exhibits include a mammoth that had been preserved by permafrost.

Moscow: Palaeontological Museum Its impressive displays include five skeletons of the big, Mongolian flesh-eating dinosaur *Tarbosaurus*.

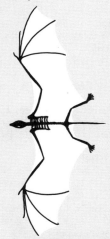

A reconstruction of the early fossil bat in a Princeton Museum

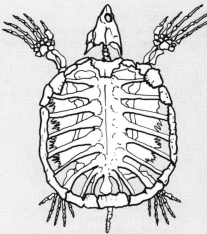

Archelon skeleton in the National Museum of Natural History, Washington, D.C.

FURTHER READING

General

Archer, M. and Clayton, G. (editors) *Vertebrate Zoogeography & Evolution in Australasia (Animals in Space & Time)* Hesperian Press, Australia 1984

Colbert, E. H. *Evolution of the Vertebrates* Wiley, 1980

McFarland, W. N.; Pough, F. H.; Cade, T. J.; and Heiser, J. B. *Vertebrate Life* Macmillan, USA 1979; Collier Macmillan, UK 1979

Moody, R. *Prehistoric World* Chartwell Books, 1980

Romer, A. S. *Vertebrate Palaeontology* University of Chicago Press, 1966

Sheehan, A. (editor) *The Prehistoric World* Galley Press, 1983

Steel, R. and Harvey, A. P. (editors) *The Encyclopaedia of Prehistoric Life* Mitchell Beazley, UK 1979; McGraw-Hill, USA 1979

Swinnerton, H. H. *Fossils* Collins, 1960

Note: News of major fossil discoveries appears in serious newspapers and in science magazines such as *Nature,* but many finds appear in specialist journals such as the *Journal of Paleontology; Paleontology;* and *Paleobiology.*

Chapter 1: Fossil Clues to Prehistoric Life

Colbert, E. H. *Wandering Lands and Animals* Dutton, USA 1973; Hutchinson, UK 1974

Fortey, R. *Fossils: The Key to the Past* Heinemann, 1982

Raup, D. M. and Stanley, S. M. *Principles of Paleontology* W. H. Freeman and Co, USA 1978

Chapter 2: Fossil Plants

Stewart, W. N. *Paleobotany and the Evolution of Plants* Cambridge University Press, 1983

Taylor, T. N. *Paleobotany* McGraw-Hill, USA 1981

Chapter 3: Fossil Invertebrates

Lehmann, U. and Hillmer, G. *Fossil Invertebrates* Cambridge University Press, 1983

Moore, R. C. et al. *Invertebrate Fossils* McGraw-Hill, USA 1952

Chapter 4: Fossil Fishes

Schultze, H.-P. *Handbook of Paleoichthyology* Fischer, West Germany 1978– (10 volumes planned)

Chapter 5: Fossil Amphibians;

Chapter 6: Fossil Reptiles

Charig, A. J. *A New Look at the Dinosaurs* Heinemann, UK 1979; Smith Publishers, USA 1979

Kemp, T. *Mammal-like Reptiles and the Origins of Mammals* Academic Press, 1982

Kuhn, O. (editor) *Encyclopedia of Paleoherpetology* Fischer, West Germany, 1969– (19 volumes planned)

Lambert, D. with the Diagram Group *Collins Guide to Dinosaurs* Collins, UK 1983; and as *A Field Guide to Dinosaurs* Avon Books, USA 1983

Swinton, W. E. *Fossil Amphibians and Reptiles* British Museum (Natural History), 1973

Chapter 7: Fossil Birds

Feduccia, A. *The Age of Birds* Harvard University Press, 1980

Swinton, W. E. *Fossil Birds* British Museum (Natural History), 1975

Chapter 8: Fossil Mammals

Kurtén, B. *The Age of Mammals* Weidenfeld and Nicolson, UK 1971; Columbia University Press, USA 1972

Clapham, F. M. (editor) *The Rise of Man* Sampson Low, 1976

Gribbin, J. and Cherfas, J. *The Monkey Puzzle,* Bodley Head, UK 1982; McGraw-Hill, USA 1983

Halstead, L. B. *The Evolution of the Mammals* Book Club Associates, 1979

Chapter 9: Records in the Rocks

Dott, Jr., R. H. et al. *Evolution of the Earth* McGraw-Hill, USA 1981

Spinar, Z. V. and Burian, Z. *Life Before Man* Thames and Hudson, UK 1972; McGraw-Hill, USA 1972

Chapter 10: Fossil Hunting

Hamilton, R. and Insole, A. N. *Finding Fossils* Penguin Books, 1977

Macfall, R. P. and Wollin, J. *Fossils for Amateurs: A Guide to Collecting and Preparing Invertebrate Fossils* Van Nostrand, USA 1972

Murray, M. *Hunting for Fossils* Macmillan, USA 1974

Rixon, A. E. *Fossil Animal Remains: Their Preparation and Their Conservation* Humanities Press, USA 1976; Athlone Press, UK 1976

INDEX